A

B

Dime Biographical Library No. 6.

THE LIFE,

AND

Military and Civic Services

OF

LIEUT.-GEN. WINFIELD SCOTT.

BY O. J. VICTOR,

Author of "THE LIFE OF GARIBALDI," &c., &c.

The life of this most remarkable of living generals is so fraught
with interest at this crisis of our national affairs as to render it a
theme eminently proper for the biographer's pen.   Alexander Hum-
boldt, after carefully studying every move in the Mexican Campaign,
pronounced him the greatest general of modern times—an opinion
endorsed quite generally in Europe.   In Great Britain he is regarded as the equal of the Duke of Wellington in all the high qualities
of command and strategy.   In this country he is so enshrined in the
hearts of the people that all eyes now instinctively turn to him.

Mr. Victor will bring to bear all the resources of his mind and
pen in the preparation of this biography, and the publishers may
safely promise a book worthy of its noble subject and romantic
theme.   It will cover the entire ground of the general's varied life
down to the present time.

Let the Youth of our land read it for its lessons of courage, faith,
patriotism, and integrity.

☞ For Sale by all News Dealers.

BEADLE AND COMPANY, Publishers,
141 William St., New York

THE

MILITARY HAND-BOOK,

AND

SOLDIER'S MANUAL OF INFORMATION.

EMBRACING

THE OFFICIAL ARTICLES OF WAR; INSTRUCTIONS TO THE
VOLUNTEER; ARMY REGULATIONS FOR CAMP AND
SERVICE; RATION AND PAY LISTS; GENERAL
RULES AND ORDERS ON ALL OCCASIONS;
HINTS ON FOOD AND ITS PREPARA-
TION; HEALTH DEPARTMENT;
WITH VALUABLE REME-
DIES AND INSTRUC-
TIONS, ETC.,

TOGETHER WITH A COMPLETE

DICTIONARY OF MILITARY TERMS.

BY LOUIS LE GRAND, M. D.

BEADLE AND COMPANY,
NEW YORK: 141 WILLIAM STREET.
LONDON: 44 PATERNOSTER ROW.

PUBLISHERS' NOTICE.

The great need of a Hand-Book of popular and useful Information for the Soldier, adapted to the exigencies of a soldier's many wants, has impelled the publishers to have prepared, at great trouble and expense, this Manual of Instruction and Information. It will be found *very complete* in all departments, and contains, beyond doubt, more truly desirable matter for the soldier and officer than any two works yet presented to the public.

The Articles of War are justly denominated "The Soldier's Gospel;" yet this is the first instance we believe in which they have been printed for general and popular circulation.

The unique and novel Dictionary of Military Terms and Science is all that any military student or soldier could ask :—it is very complete.

The publishers trust that the low price at which this valuable Hand-Book is sold will make it acceptable to all, and render it, what it was designed to be,

A SOLDIER'S COMPANION AND GUIDE.

NEW YORK, June 10th, 1861.

INTRODUCTORY.

WE have sought, in the preparation of this work, to supply a want painfully felt by officers as well as soldiers, of a Hand and Text Book for the Soldier; embracing, besides the usual instruction upon the routine and duties of service-life, all·the information in regard to Food and its varied preparation—Health, and how to retain it—Special Directions for Special Emergencies, etc., etc. Such a manual has not been, singularly enough, yet produced; and, in essaying the task of supplying the want, we have studied to leave nothing unsaid which could prove useful and necessary to the instruction and comfort of the soldier.

We find, in consulting authorities, considerable discrepancy in the statements of different authors upon the same point. Some of these are *glaringly wrong*, and are designed to prove a source of disappointment to soldiers and of annoyance to officers. Thus, some parties state that, previous to being called into government service, the infantry volunteer and militia privates get twenty-one dollars per month, pay. They only receive eleven dollars. The Oath of Allegiance has been misprinted. The Pay Department table has been incorrectly reproduced. The Rations list has been *materially misstated* in two recent "cheap," and *professedly official* works. So the record runs.

The addition to this work of the Articles of War, is to meet the demand of the hour. No work yet before the public, other than the expensive "Army Regulations," has printed this document at length—a document as valuable to every soldier as the Constitution and Laws of our Country are to every citizen. We have used the official copy, and suggested, in notes, such modifications as have been made of the Original Articles.

The Dictionary of Military Terms and Science is, so far as we are aware, entirely new, in this country. We have sought to render it thorough and exhaustive, making use of the best English authorities for its basis.

In the preparation of the Culinary Department, we have used Soyer's celebrated "Army Recipes," so far as we deemed them available, but have made important additions to his list of dishes. This department will be found as explicit and *adaptive* as could be desired.

The Health Department embraces what is best of recent information on the subject of the condition, health and comfort of the soldier. A reference to the index will show what valuable data is now made available for the exigencies of military life.

The Reference Index will materially aid in quickly finding any information, on any required point.

Trusting our labor may prove acceptable to the soldier, and serve him as a Guide and Companion, it is submitted, in the earnest hope that it may, in some small degree, at least, contribute to his well-being at this great crisis in the affairs of our beloved country.

L. Le G.

New York, *June 1st*, 1861.

GENERAL INDEX.

REFERENCE INDEX.

THE MILITARY HAND-BOOK.

ARTICLES OF WAR

FOR THE GOVERNMENT OF THE ARMY OF THE UNITED STATES.

[OFFICIAL COPY.]

ARTICLE 1. Every officer now in the army of the United States shall, in six months from the passing of this act, and every officer who shall hereafter be appointed shall, before he enters on the duties of his office, subscribe these rules and regulations.

ART. 2. It is earnestly recommended to all officers and soldiers diligently to attend divine service; and all officers who shall behave indecently or irreverently at any place of divine worship shall, if commissioned officers, be brought before a general court-martial, there to be publicly and severely reprimanded by the president; if non-commissioned officers or soldiers, every person so offending shall, for his first offense, forfeit one-sixth of a dollar, to be deducted out of his next pay; for the second offense, he shall not only forfeit a like sum, but be confined twenty-four hours; and for every like offense, shall suffer and pay in like manner; which money, so forfeited, shall be applied, by the captain or senior officer of the troop or company, to the use of the sick soldiers of the company or troop to which the offender belongs.

ART. 3. Any non-commissioned officer or soldier who shall use any profane oath or execration, shall incur the penalties expressed in the foregoing article; and a commissioned officer shall forfeit and pay, for each and every such offense, one dollar, to be applied as in the preceding article.

ART. 4. Every chaplain commissioned in the army or armies of the United States, who shall absent himself from the duties assigned him (excepting in cases of sickness or leave of absence), shall, on conviction thereof before a court-martial, be fined not exceeding one month's pay, besides the loss of his pay during his absence; or be discharged, as the said court-martial shall judge proper.

ART. 5. Any officer or soldier who shall use contemptuous or disrespectful words against the President of the United States, against the Vice-President thereof, against the Congress of the United States, or against the Chief Magistrate or Legislature of any of the United States, in which he may be quartered, if a commissioned officer, shall be cashiered, or otherwise punished, as a court-martial shall direct; if a non-commissioned officer or soldier, he shall suffer such punishment as shall be inflicted on him by the sentence of a court-martial.

ART. 6. Any officer or soldier who shall behave himself with contempt or disrespect toward his commanding officer, shall be punished, according to the nature of his offense, by the judgment of a court-martial.

ART. 7. Any officer or soldier who shall begin, excite, cause, or join in, any mutiny or sedition, in any troop or company in the service of the United States, or in any party, post, detachment, or guard, shall suffer death, or such other punishment as by a court-martial shall be inflicted.

ART. 8. Any officer, non-commissioned officer, or soldier, who, being present at any mutiny or sedition, does not use his utmost endeavor to suppress the same, or, coming to the knowledge of any intended mutiny, does not, without delay, give information thereof to his commanding officer, shall be punished by the sentence of a court-martial with death, or otherwise, according to the nature of his offense.

ART. 9. Any officer or soldier who shall strike his superior officer, or draw or lift up any weapon, or offer any violence against him, being in the execution of his office, on any pretense whatsoever, or shall disobey any lawful command of his superior officer, shall suffer death, or such other punishment as shall, according to the nature of his offense, be inflicted upon him by the sentence of a court-martial.

ART. 10. Every non-commissioned officer or soldier, who shall enlist himself in the service of the United States, shall, at the time of his so enlisting, or within six days afterward, have the Articles for the government of the armies of the United States read to him, and shall, by the officer who enlisted him, or by the commanding officer of the troop or company into which he was enlisted, be taken before the next justice of the peace, or chief magistrate of any city or town corporate, not being an officer of the army, or where recourse can not be had to the civil magistrate, before the judge advocate, and in his presence shall take the following oath or affirmation: "I, A. B., do solemnly swear or affirm (as the case may be), that I will bear true allegiance to the United States of America, and that I will serve them honestly and faithfully against all their enemies or opposers whatsoever; and observe and obey the orders of the President of the United States, and the orders of the officers appointed over me, according to the Rules and Articles for the government of the armies of the United States." Which justice, magistrate, or judge advocate is to give to the officer a certificate, signifying that the man enlisted did take the said oath or affirmation.

ART. 11. After a non-commissioned officer or soldier shall have been duly enlisted and sworn, he shall not be dismissed the service without a discharge in writing; and no discharge granted to him shall be sufficient which is not signed by a field officer of the regiment to which he belongs, or commanding officer, where no field officer of the regiment is present; and no discharge shall be given to a non-commissioned officer or soldier before his term of service has expired, but by order of the President, the Secretary of War, the commanding officer of a department, or the sentence of a general court-martial; nor shall a commissioned officer be discharged the service but by order of the President of the United States, or by sentence of a general court-martial.

ART. 12. Every colonel, or other officer commanding a regiment, troop, or company, and actually quartered with it, may give furloughs to non-commissioned officers or soldiers, in such numbers, and for so long a time, as he shall judge to be most consistent with the good of the service; and a captain, or other inferior officer, commanding a troop or company, or in any garrison, fort, or barrack of the United States (his field officer being absent), may give furloughs to non-commissioned officers or soldiers, for a time not exceeding twenty days in six months, but not to more than two persons to be absent at the same time, excepting some extraordinary occasion should require it.

ART. 13. At every muster, the commanding officer of each regiment, troop, or company, there present, shall give to the commissary of musters, or other officer who musters the said regiment, troop, or company, certificates signed by himself, signifying how long such officers, as shall not appear at the said muster, have been absent, and the reason of their absence. In like manner, the commanding officer of every troop or company shall give certificates, signifying the reasons of the absence of the non-commissioned officers and private soldiers; which reasons and time of absence shall be inserted in the muster-rolls, opposite the names of the respective absent officers and soldiers. The certificates shall, together with the muster-rolls, be remitted by the commissary of musters, or other officer mustering, to the Department of War, as speedily as the distance of the place will admit.

ART. 14. Every officer who shall be convicted before a general court-martial of having signed a false certificate relating to the absence of either officer or private soldier, or relative to his or their pay, shall be cashiered.

ART. 15. Every officer who shall knowingly make a false muster of man or horse, and every officer or commissary of musters who shall willingly sign, direct, or allow the signing of muster-rolls, wherein such false muster is contained, shall, upon proof made thereof, by two witnesses, before a general court-martial, be cashiered, and shall be thereby utterly disabled to have or hold any office or employment in the service of the United States.

ART. 16. Any commissary of musters, or other officer, who shall be convicted of having taken money, or other thing, by way of gratification, on mustering any regiment, troop, or company, or on signing muster-rolls, shall be displaced from his office, and shall be thereby utterly disabled to have or hold any office or employment in the service of the United States.

ART. 17. Any officer who shall presume to muster a person as a soldier who is not a soldier, shall be deemed guilty of having made a false muster, and shall suffer accordingly.

ART. 18. Every officer who shall knowingly make a false return to the Department of War, or to any of his superior officers, authorized to call for such returns, of the state of the regiment, troop, or company, or garrison, under his command; or of the arms, ammunition, clothing, or other stores thereunto belonging, shall, on conviction thereof before a court-martial, be cashiered.

ART. 19. The commanding officer of every regiment, troop, or independent company, or garrison, of the United States, shall, in the beginning of every month, remit, through the proper channels, to the Department of War, an exact return of the regiment, troop, independent company, or garrison, under his command, specifying the names of the officers then absent from their posts, with the reasons for and the time of their absence. And any officer who shall be convicted of having, through neglect or design, omitted sending such returns, shall be punished, according to the nature of his crime, by the judgment of a general court-martial.

ART. 20. All officers and soldiers who have received pay, or have been duly enlisted in the service of the United States, and shall be convicted of having deserted the same, shall suffer death, or such other punishment as, by sentence of a court-martial, shall be inflicted.*

ART. 21. Any non-commissioned officer or soldier who shall, without leave from his commanding officer, absent himself from his troop,

---

* Modified (by act of 29th May, 1830) to the effect that no officer or soldier in the army shall be subject to the punishment of death for desertion in time of peace.

company, or detachment, shall, upon being convicted thereof, be punished according to the nature of his offense, at the discretion of a court-martial.

ART. 22. No non-commissioned officer or soldier shall enlist himself in any other regiment, troop or company, without a regular discharge from the regiment, troop, or company in which he last served, on the penalty of being reputed a deserter, and suffering accordingly. And in case any officer shall knowingly receive and entertain such non-commissioned officer or soldier, or shall not, after his being discovered to be a deserter, immediately confine him, and give notice thereof to the corps in which he last served, the said officer shall, by a court-martial, be cashiered.

ART. 23. Any officer or soldier who shall be convicted of having advised or persuaded any other officer or soldier to desert the service of the United States, shall suffer death, or such other punishment as shall be inflicted upon him by the sentence of a court-martial.

ART. 24. No officer or soldier shall use any reproachful or provoking speeches or gestures to another, upon pain, if an officer, of being put in arrest; if a soldier, confined, and of asking pardon of the party offended, in the presence of his commanding officer.

ART. 25. No officer or soldier shall send a challenge to another officer or soldier, to fight a duel, or accept a challenge if sent, upon pain, if a commissioned officer, of being cashiered; if a non-commissioned officer or soldier, of suffering corporeal punishment, at the discretion of a court-martial.

ART. 26. If any commissioned or non-commissioned officer commanding a guard shall knowingly or willingly suffer any person whatsoever to go forth to fight a duel, he shall be punished as a challenger; and all seconds, promoters, and carriers of challenges, in order to duels, shall be deemed principals, and be punished accordingly. And it shall be the duty of every officer commanding an army, regiment, company, post, or detachment, who is knowing to a challenge being given or accepted by any officer, non-commissioned officer, or soldier, under his command, or has reason to believe the same to be the case, immediately to arrest and bring to trial such offenders.

ART. 27. All officers, of what condition soever, have power to part and quell all quarrels, frays, and disorders, though the persons concerned should belong to another regiment, troop, or company; and either to order officers into arrest, or non-commissioned officers or soldiers into confinement, until their proper superior officers shall be acquainted therewith; and whosoever shall refuse to obey such officer (though of an inferior rank), or shall draw his sword upon him, shall be punished at the discretion of a general court-martial.

Art. 28. Any officer or soldier who shall upbraid another for refusing a challenge, shall himself be punished as a challenger; and all officers and soldiers are hereby discharged from any disgrace or opinion of disadvantage which might arise from their having refused to accept of challenges, as they will only have acted in obedience to the laws, and done their duty as good soldiers who subject themselves to discipline.

Art. 29. No sutler shall be permitted to sell any kind of liquors or victuals, or to keep their houses or shops open for the entertainment of soldiers, after nine at night, or before the beating of the reveille, or upon Sundays, during divine service or sermon, on the penalty of being dismissed from all future sutling.

Art. 30. All officers commanding in the field, forts, barracks, or garrisons of the United States, are hereby required to see that the persons permitted to suttle shall supply the soldiers with good and wholesome provisions, or other articles, at a reasonable price, as they shall be answerable for their neglect.

Art. 31. No officer commanding in any of the garrisons, forts, or barracks of the United States, shall exact exorbitant prices for houses or stalls, let out to sutlers, or connive at the like exactions in others; nor by his own authority, and for his private advantage, lay any duty or imposition upon, or be interested in the sale of any victuals, liquors, or other necessaries of life brought into the garrison, fort, or barracks, for the use of the soldiers, on the penalty of being discharged from the service.

Art. 32. Every officer commanding in quarters, garrisons, or on the march, shall keep good order, and, to the utmost of his power, redress all abuses or disorders which may be committed by any officer or soldier under his command; if, upon complaint made to him of officers or soldiers beating or otherwise ill-treating any person, or disturbing fairs or markets, or of committing any kind of riots, to the disquieting of the citizens of the United States, he, the said commander, who shall refuse or omit to see justice done to the offender or offenders, and reparation made to the party or parties injured, as far as part of the offender's pay shall enable him or them, shall, upon proof thereof, be cashiered, or otherwise punished, as a general court-martial shall direct.

Art. 33. When any commissioned officer or soldier shall be accused of a capital crime, or of having used violence, or committed any offense against the person or property of any citizen of any of the United States, such as is punishable by the known laws of the land, the commanding officer and officers of every regiment, troop or company, to which the person or persons so accused shall belong, are hereby required, upon application duly made by, or in behalf of the party or parties injured, to use their utmost endeavors to deliver over such accused person or persons to the civil magistrate, and likewise to be aiding and assisting to the officers of justice in apprehending and securing the person or persons so accused, in order to bring him or them to trial. If any commanding officer or officers shall willfully neglect, or shall refuse, upon the application aforesaid, to deliver over such accused person or persons to the civil magistrates, or to be aiding and assisting to the officers of justice in apprehending such person or persons, the officer or officers so offending shall be cashiered.

Art. 34. If any officer shall think himself wronged by his colonel, or the commanding officer of the regiment, and shall, upon due application being made to him, be refused redress, he may complain to the general commanding in the State or Territory where such regiment shall be stationed, in order to obtain justice; who is hereby required to examine into said complaint, and take proper measures for redressing the wrong complained of, and transmit, as soon as possible, to the Department of War, a true state of such complaint, with the proceedings had thereon.

Art. 35. If any inferior officer or soldier shall think himself wronged by his captain or other officer, he is to complain thereof to the commanding officer of the regiment, who is hereby required to summon a regimental court-martial, for the doing justice to the complainant; from which regimental court-martial either party may, if he thinks himself still aggrieved, appeal to a general court martial. But if, upon a second hearing, the appeal shall appear vexatious and groundless, the person so appealing shall be punished at the discretion of the said court-martial.

Art. 36. Any commissioned officer, store-keeper, or commissary, who shall be convicted at a general court-martial of having sold, without a proper order for that purpose, embezzled, misapplied, or willfully, or through neglect, suffered any of the provisions, forage, arms, clothing, ammunition, or other military stores belonging to the United States to be spoiled or damaged, shall, at his own expense, make good the loss or damage, and shall, moreover, forfeit all his pay, and be dismissed from the service.

Art. 37. Any non-commissioned officer or soldier who shall be convicted at a regimental court-martial of having sold, or designedly or through neglect, wasted the ammunition delivered out to him, to be employed in the service of the United States, shall be punished at the discretion of such court.

Art. 38. Every non-commissioned officer or soldier who shall be convicted before a court-martial of having sold, lost or spoiled, through neglect, his horse, arms, clothes, or accouterments, shall undergo with weekly stoppages (not exceeding the half of his pay) as such court-martial shall judge sufficient, for repairing the loss or damage; and shall suffer confinement, or such other corporeal punishment as his crime shall deserve.

Art. 39. Every officer who shall be convicted before a court-martial of having embezzled or misapplied any money with which he may have been intrusted, for the payment of the men under his command, or for enlisting men into the service, or for other purposes, if a commissioned officer, shall be cashiered, and compelled to refund the money; if a non-commissioned officer, shall be reduced to the ranks, be put under stoppages until the money be made good, and suffer such corporeal punishment as such court-martial shall direct.

Art. 40. Every captain of a troop or company is charged with the arms, accouterments, ammunition, clothing, or other warlike stores belonging to the troop or company under his command, which he is to be accountable for to his colonel in case of their being lost, spoiled, or damaged, not by unavoidable accidents, or on actual service.

Art. 41. All non-commissioned officers and soldiers who shall be found one mile from the camp without leave, in writing, from their commanding officer, shall suffer such punishment as shall be inflicted upon them by the sentence of a court-martial.

Art. 42. No officer or soldier shall lie out of his quarters, garrison, or camp without leave from his superior officer, upon penalty of being punished according to the nature of his offense, by the sentence of a court-martial.

Art. 43. Every non-commissioned officer and soldier shall retire to his quarters or tent at the beating of the retreat; in default of which he shall be punished according to the nature of his offense.

Art. 44. No officer, non-commissioned officer, or soldier shall fail in repairing, at the time fixed, to the place of parade, of exercise, or other rendezvous appointed by his commanding officer, if not prevented by sickness or some other evident necessity, or shall go from the said place of rendezvous without leave from his commanding officer, before he shall be regularly dismissed or relieved, on the penalty of being punished, according to the nature of his offense, by the sentence of a court-martial.

Art. 45. Any commissioned officer who shall be found drunk on his guard, party, or other duty, shall be cashiered. Any non-commissioned officer or soldier so offending shall suffer such corporeal punishment as shall be inflicted by the sentence of a court-martial.

Art. 46. Any sentinel who shall be found sleeping upon his post, or shall leave it before he shall be regularly relieved, shall suffer death, or such other punishment as shall be inflicted by the sentence of a court-martial.

ART. 47. No soldier belonging to any regiment, troop or company shall hire another to do his duty for him, or be excused from duty but in cases of sickness, disability, or leave of absence; and every such soldier found guilty of hiring his duty, as also the party so hired to do another's duty, shall be punished at the discretion of a regimental court-martial.

ART. 48. And every non-commissioned officer conniving at such hiring of duty aforesaid, shall be reduced; and every commissioned officer knowing and allowing such ill-practices in the service, shall be punished by the judgment of a general court-martial.

ART. 49. Any officer belonging to the service of the United States, who, by discharging of firearms, drawing of swords, beating of drums, or by any other means whatsoever, shall occasion false alarms in camp, garrison, or quarters, shall suffer death, or such other punishment as shall be ordered by the sentence of a general court-martial.

ART. 50. Any officer or soldier who shall, without urgent necessity, or without the leave of his superior officer, quit his guard, platoon, or division, shall be punished, according to the nature of his offense, by the sentence of a court-martial.

ART. 51. No officer or soldier shall do violence to any person who brings provisions or other necessaries to the camp, garrison, or quarters of the forces of the United States, employed in any parts out of the said States, upon pain of death, or such other punishment as a court-martial shall direct.

ART. 52. Any officer or soldier who shall misbehave himself before the enemy, run away, or shamefully abandon any fort, post, or guard which he or they may be commanded to defend, or speak words inducing others to do the like, or shall cast away his arms and ammunition, or who shall quit his post or colors to plunder and pillage, every such offender, being duly convicted thereof, shall suffer death, or such other punishment as shall be ordered by the sentence of a general court-martial.

ART. 53. Any person belonging to the armies of the United States who shall make known the watchword to any person who is not entitled to receive it according to the rules and discipline of war, or shall presume to give a parole or watchword different from what he received, shall suffer death, or such other punishment as shall be ordered by the sentence of a general court-martial.

ART. 54. All officers and soldiers are to behave themselves orderly in quarters and on their march; and whoever shall commit any waste or spoil, either in walks of trees, parks, warrens, fish-ponds, houses, or gardens, corn-fields, inclosures of meadows, or shall maliciously destroy any property whatsoever belonging to the inhabitants of the United States, unless by order of the then commander-in-chief of the armies of the said States, shall (besides such penalties as they are liable to by law), be punished according to the nature and degree of the offense. by the judgment of a regimental or general court-martial.

Art. 55. Whosoever, belonging to the armies of the United States in foreign parts, shall force a safeguard, shall suffer death.

Art. 56. Whosoever shall relieve the enemy with money, victuals, or ammunition, or shall knowingly harbor or protect an enemy, shall suffer death, or such other punishment as shall be ordered by the sentence of a court-martial.

Art. 57. Whosoever shall be convicted of holding correspondence with, or giving intelligence to, the enemy, either directly or indirectly, shall suffer death, or such other punishment as shall be ordered by the sentence of a court-martial.

Art. 58. All public stores taken in the enemy's camp, towns, forts, or magazines, whether of artillery, ammunition, clothing, forage or provisions, shall be secured for the service of the United States; for the neglect of which the commanding officer is to be answerable.

Art. 59. If any commander of any garrison, fortress or post shall be compelled, by the officers and soldiers under his command, to give up to the enemy, or to abandon it, the commissioned officers, non-commissioned officers, or soldiers who shall be convicted of having so offended, shall suffer death, or such other punishment as shall be inflicted upon them by the sentence of a court-martial.

Art. 60. All sutlers and retainers to the camp, and all persons whatsoever, serving with the armies of the United States in the field, though not enlisted soldiers, are to be subject to orders, according to the rules and discipline of war.

Art. 61. Officers having brevets or commissions of a prior date to those of the regiment in which they serve, may take place in courts-martial and on detachments, when composed of different corps, according to the ranks given them in their brevets or dates of their former commissions; but in the regiment, troop, or company to which such officers belong, they shall do duty and take rank both in courts-martial and on detachments, which shall be composed of their own corps, according to the commissions by which they are mustered in the said corps.

Art. 62. If, upon marches, guards, or in quarters, different corps of the army shall happen to join, or do duty together, the officer highest in rank of the line of the army, marine corps, or militia, by commission, there on duty or in quarters, shall command the whole, and give orders for what is needful to the service, unless otherwise specially directed by the President of the United States, according to the nature of the case.

ART. 63. The functions of the engineers being generally confined to the most elevated branch of military science, they are not to assume, nor are they subject to be ordered on any duty beyond the line of their immediate profession, except by the special order of the President of the United States ; but they are to receive every mark of respect to which their rank in the army may entitle them respectively, and are liable to be transferred, at the discretion of the President, from one corps to another, regard being paid to rank.

ART. 64. General courts-martial may consist of any number of commissioned officers, from five to thirteen, inclusively ; but they shall not consist of less than thirteen where that number can be convened without manifest injury to the service.

ART. 65.* Any general officer commanding an army, or colonel commanding a separate department, may appoint general courts-martial whenever necessary. But no sentence of a court-martial shall be carried into execution until after the whole proceedings shall have been laid before the officer ordering the same, or the officer commanding the troops for the time being; neither shall any sentence of a general court-martial, in the time of peace, extending to the loss of life, or the dismission of a commissioned officer, or which shall, either in time of peace or war, respect a general officer, be carried into execution, until after the whole proceedings shall have been transmitted to the Secretary of War, to be laid before the President of the United States for his confirmation or disapproval, and orders in the case. All other sentences may be confirmed and executed by the officer ordering the court to assemble, or the commanding officer for the time being, as the case may be.

ART. 66. Every officer commanding a regiment or corps may appoint for his own regiment or corps, courts-martial, to consist of three commissioned officers, for the trial and punishment of offenses not capital, and decide upon their sentences. For the same purpose, all officers commanding any of the garrisons, forts, barracks, or other places where the troops consist of different corps, may assemble courts-martial, to consist of three commissioned officers, and decide upon their sentences.

ART. 67. No garrison or regimental court-martial shall have the power to try capital cases or commissioned officers; neither shall they inflict a fine exceeding one month's pay, nor imprison, nor put to hard

---

* Modified by act of 29th May, 1830, to the effect that the proceedings and sentence shall be *sent direct* to the Secretary, etc.

labor, any non-commissioned officer or soldier for a longer time than one month.

Art. 68. Whenever it may be found convenient and necessary to the public service, the officers of the marines shall be associated with the officers of the land forces, for the purpose of holding courts-martial, and trying offenders belonging to either; and, in such cases, the orders of the senior officer of either corps who may be present and duly author ized, shall be received and obeyed.

Art. 69. The judge advocate, or some person deputed by him, or by the general, or officer commanding the army, detachment, or garrison, shall prosecute in the name of the United States, but shall so far consider himself as counsel for the prisoner, after the said prisoner shall have made his plea, as to object to any leading question to any of the witnesses, or any question to the prisoner, the answer to which might tend to criminate himself; and administer to each member of the court, before they proceed upon any trial, the following oath, which shall also be taken by all members of the regimental and garrison courts-martial : "You, A. B., do swear that you will well and truly try and determine, according to evidence, the matter now before you, between the United States of America and the prisoner to be tried, and that you will duly administer justice, according to the provisions of 'An act establishing Rules and Articles for the government of the armies of the United States,' without partiality, favor, or affection ; and if any doubt should arise, not explained by said Articles, according to your conscience, the best of your understanding, and the custom of war in like cases; and you do further swear that you will not divulge the sentence of the court until it shall be published by the proper authority; neither will you disclose or discover the vote or opinion of any particular member of the court-martial, unless required to give evidence thereof, as a witness by a court of justice, in a due course of law. So help you God." And as soon as the said oath shall have been administered to the respective members, the president of the court shall administer to the judge advocate, or person officiating as such, an oath in the following words : "You, A. B., do swear, that you will not disclose or discover the vote or opinion of any particular member of the court-martial, unless required to give evidence thereof, as a witness, by a court of justice, in due course of law; nor divulge the sentence of the court to any but the proper authority, until it shall be duly disclosed by the same. So help you God."

Art. 70. When a prisoner, arraigned before a general court-martial, shall, from obstinacy and deliberate design, stand mute, or answer foreign to the purpose, the court may proceed to trial and judgment as if the prisoner had regularly pleaded not guilty.

Art. 71. When a member shall be challenged by a prisoner, he must state his cause of challenge, of which the court shall, after due deliberation, determine the relevancy or validity, and decide accordingly ; and no challenge to more than one member at a time shall be received by the court.

Art. 72. All the members of a court-martial are to behave with decency and calmness ; and in giving their votes are to begin with the youngest in commission.

Art. 73. All persons who give evidence before a court-martial are to be examined on oath or affirmation, in the following form : " You swear, or affirm (as the case may be), the evidence you shall give in the cause now in hearing shall be the the truth, the whole truth, and nothing but the truth. So help you God."

Art. 74. On the trials of cases not capital, before courts-martial, the deposition of witnesses, not in the line or staff of the army, may be taken before some justice of the peace, and read in evidence ; provided the prosecutor and person accused, are present at the taking the same, or are duly notified thereof.

Art. 75. No officer shall be tried but by a general court-martial, nor by officers of an inferior rank, if it can be avoided. Nor shall any proceedings of trials be carried on, excepting between the hours of eight in the morning and three in the afternoon, excepting in cases which, in the opinion of the officer appointing the court-martial, require immediate example.

Art. 76. No person whatsover shall use any menacing words, signs, or gestures, in presence of a court-martial, or shall cause any disorder or riot, or disturb their proceedings, on the penalty of being punished at the discretion of the said court-martial.

Art. 77. Whenever any officer shall be charged with a crime, he shall be arrested and confined in his barracks, quarters, or tent, and deprived of his sword by the commanding officer. And any officer who shall leave his confinement before he shall be set at liberty by his commanding officer, or by a superior officer, shall be cashiered.

Art. 78. Non-commissioned officers and soldiers, charged with crimes, shall be confined until tried by a court-martial, or released by proper authority.

Art. 79. No officer or soldier who shall be put in arrest shall continue in confinement more than eight days, or until such time as a court-martial can be assembled.

Art. 80. No officer commanding a guard, or provost marshal, shall refuse to receive or keep any prisoner committed to his charge by an officer belonging to the forces of the United States ; provided the officer committing shall, at the same time, deliver an account in writing, signed by himself, of the crime with which the said prisoner is charged.

ART. 81. No officer commanding a guard, or provost marshal, shall presume to release any person committed to his charge without proper authority for so doing, nor shall he suffer any person to escape on the penalty of being punished for it by the sentence of a court-martial.

ART. 82. Every officer or provost marshal, to whose charge prisoners shall be committed, shall, within twenty-four hours after such commitment, or as soon as he shall be relieved from his guard, make report in writing, to the commanding officer, of their names, their crimes, and the names of the officers who committed them, on the penalty of being punished for disobedience or neglect, at the discretion of a court-martial.

ART. 83. Any commissioned officer convicted before a general court-martial of conduct unbecoming an officer and a gentleman, shall be dismissed the service.

ART. 84. In cases where a court-martial may think it proper to sentence a commissioned officer to be suspended from command, they shall have power also to suspend his pay and emoluments for the same time, according to the nature and heinousness of the offense.

ART. 85. In all cases where a commissioned officer is cashiered for cowardice or fraud, it shall be added in the sentence, that the crime, name, and place of abode, and punishment of the delinquent, be published in the newspapers in and about the camp, and of the particular State from which the offender came, or where he usually resides; after which it shall be deemed scandalous for an officer to associate with him.

ART. 86. The commanding officer of any post or detachment, in which there shall not be a number of officers adequate to form a general court-martial, shall, in cases which require the cognizance of such a court, report to the commanding officer of the department, who shall order a court to be assembled at the nearest post or department, and the party accused, with necessary witnesses, to be transported to the place where the said court shall be assembled.

ART. 87.* No person shall be sentenced to suffer death but by the concurrence of two-thirds of the members of a general court-martial, nor except in the cases herein expressly mentioned; *nor shall more than fifty lashes be inflicted on any offender, at the discretion of a*

---

* So much of these rules and articles as authorizes the infliction of corporeal punishment by stripes or lashes, was specially repealed by act of 16th May, 1812. By act of 2d March, 1833, the repealing act was repealed, so far as it applied to the crime of desertion, which, of course, revived the punishment by lashes for that offense.

court-martial; and no officer, non-commissioned officer, soldier or follower of the army, shall be tried a second time for the same offense.

ART. 88. No person shall be liable to be tried and punished by a general court-martial for any offense which shall appear to have been committed more than two years before the issuing of the order for such trial, unless the person, by reason of having absented himself or some other manifest impediment, shall not have been amenable to justice within that period.

ART. 89. Every officer authorized to order a general court-martial shall have power to pardon or mitigate any punishment ordered by such court, except the sentence of death, or of cashiering an officer ; which, in the cases where he has authority (by Article 65) to carry them into execution, he may suspend, until the pleasure of the President of the United States can be known; which suspension, together with copies of the proceedings of the court-martial, the said officer shall immediately transmit to the President for his determination. And the colonel or commanding officer of the regiment or garrison where any regimental or garrison court-martial shall be held, may pardon or mitigate any punishment ordered by such court to be inflicted.

ART. 90. Every judge advocate, or person officiating as such, at any general court-martial, shall transmit, with as much expedition as the opportunity of time and distance of place can admit, the original proceedings and sentence of such court-martial to the Secretary of War; which said original proceedings and sentence shall be carefully kept and preserved in the office of said Secretary, to the end that the persons entitled thereto may be enabled, upon application to the said office, to obtain copies thereof. The party tried by any general court-martial shall, upon demand thereof, made by himself, or by any person or persons in his behalf, be entitled to a copy of the sentence and proceedings of such court-martial.

ART. 91. In cases where the general, or commanding officer may order a court of inquiry to examine into the nature of any transaction, accusation, or imputation against any officer or soldier, the said court shall consist of one or more officers, not exceeding three, and a judge advocate, or other suitable person, as a recorder, to reduce the proceedings and evidence to writing; all of whom shall be sworn to the faithful performance of their duty. This court shall have the same power to summon witnesses as a court-martial, and to examine them on oath. But they shall not give their opinion on the merits of the case, excepting they shall be thereto specially required. The parties accused shall also be permitted to cross-examine and interrogate the witnesses, so as to investigate fully the circumstances in the question.

ART. 92. The proceedings of a court of inquiry must be authenticated by the signature of the recorder and the president, and delivered to the commanding officer, and the said proceedings may be admitted as evidence by a court-martial, in cases not capital, or extending to the dismission of an officer, provided that the circumstances are such that oral testimony can not be obtained. But as courts of inquiry may be perverted to dishonorable purposes, and may be considered as engines of destruction to military merit, in the hands of weak and envious commandants, they are hereby prohibited, unless directed by the President of the United States, or demanded by the accused.

ART. 93. The judge advocate or recorder shall administer to the members the following oath : " You shall well and truly examine and inquire, according to your evidence, into the matter now before you, without partiality, favor, affection, prejudice, or hope of reward. So help you God." After which the president shall administer to the judge advocate or recorder the following oath : "You, A. B., do swear that you will, according to your best abilities, accurately and impartially record the proceedings of the court, and the evidence to be given in the case in hearing. So help you God." The witnesses shall take the same oath as witnesses sworn before a court-martial.

ART. 94. When any commissioned officer shall die or be killed in the service of the United States, the major of the regiment, or the officer doing the major's duty in his absence, or in any post or garrison, the second officer in command, or the assistant military agent, shall immediately secure all his effects or equipage, then in camp or quarters, and shall make an inventory thereof, and forthwith transmit the same to the office of the Department of War, to the end that his executors or administrators may receive the same.

ART. 95. When any non-commissioned officer or soldier shall die, or be killed in the service of the United States, the then commanding officer of the troop or company shall, in the presence of two other commissioned officers, take an account of what effects he died possessed of, above his arms and accouterments, and transmit the same to the office of the Department of War, which said effects are to be accounted for, and paid to the representatives of such deceased non-commissioned officer or soldier. And in case any of the officers, so authorized to take care of the effects of deceased officers and soldiers, should, before they have accounted to their representatives for the same, have occasion to leave the regiment or post, by preferment or otherwise, they shall, before they be permitted to quit the same, deposit in the hands of the commanding officer, or of the assistant military agent, all the effects of such deceased non-commissioned officers and soldiers, in order that the same may be secured for, and paid to their respective representatives.

Art. 96. All officers, conductors, gunners, matrosses, drivers, or other persons whatsoever, receiving pay or hire in the service of the artillery, or corps of engineers of the United States, shall be governed by the aforesaid rules and articles, and shall be subject to be tried by courts-martial, in like manner with the officers and soldiers of the other troops in the service of the United States.

Art. 97. The officers and soldiers of any troops whether militia or others, being mustered and in pay of the United States, shall, at all times and in all places, when joined, or acting in conjunction with the regular forces of the United States, be governed by these rules and articles of war, and shall be subject to be tried by courts-martial, in like manner with the officers and soldiers in the regular forces ; save only that such courts-martial shall be composed entirely of militia officers.

Art. 98. All officers serving by commission from the authority of any particular State, shall, on all detachments, courts-martial, or other duty wherein they may be employed in conjunction with the regular forces of the United States, take rank next after all officers of the like grade in said regular forces, notwithstanding the commissions of such militia or State officers may be elder than the commissions of the officers of the regular forces of the United States.

Art. 99. All crimes not capital, and all disorders and neglects which officers and soldiers may be guilty of, to the prejudice of good order and military discipline, though not mentioned in the foregoing articles of war, are to be taken cognizance of by a general or regimental court-martial, according to the nature and degree of the offense, and be punished at their discretion.

Art. 100. The President of the United States shall have power to prescribe the uniform of the army.

Art. 101. The foregoing articles are to be read and published, once in every six months, to every garrison, regiment, troop, or company, mustered, or to be mustered, in the service of the United States, and are to be duly observed and obeyed by all officers and soldiers who are, or shall be, in said service.

Sec. 2. *And be it further enacted,* That in time of war, all persons not citizens of, or owing allegiance to, the United States of America, who shall be found lurking as spies in or about the fortifications or encampments of the armies of the United States, or any of them, shall suffer death, according to the law and usage of nations, by sentence of a general court-martial.

Sec. 3. *And be it further enacted,* That the rules and regulations by which the armies of the United States have heretofore been governed, and the resolves of Congress thereunto annexed, and respecting the same, shall henceforth be void and of no effect, except so far as may relate to any transactions under them prior to the promulgation of this act, at the several posts and garrisons respectively, occupied by any part of the army of the United States. [APPROVED, April 10 1806.]

PERSONAL HINTS TO VOLUNTEERS.

A VOLUNTEER has his choice of regiments, and, therefore, can, to a great degree, choose his associations. At the best, a soldier's life is one calculated to test a man's moral as well as physical qualities ; hence, if a person is to embark in the service, he should earnestly strive to obtain a place in those regiments or companies which reject all "hard cases" and men of vicious habits. Otherwise, he will be annoyed by fellowship with creatures whom he must despise, and may be subjected to mortification, if not to actual disgrace from their bad conduct.

"Going into the tented field" really means going to privation ; to danger in several shapes ; to constant self-denial ; to a kind of slavery where your will and wish are both totally ignored in the will and wish of your *superiors ;* therefore, it behooves the volunteer to weigh the matter well in his mind that he may not regret his step, nor ever be tempted to grumble at his duty, much less to desert. One who flinches from duty, or complains at privations, is not fit for a soldier's great trust ; while a deserter is one of the most despised of men, even by those to whose protection he flees.

The "Zouave" mania is becoming, in plain words, over-done. Some men appear to think that, to be a good soldier, a man must needs look as much as possible like a Turk, who dresses with far more regard to a peacock display of colors than of utility or good taste. That this *is* true, let any one compare the solid, sober gray uniform, or the modest and sensible blue, with the flaunting firey red " bag " breeches, the uncouth jacket and senseless gew-gaw trimmings of the " Zouave," and we will venture to say he can have but one opinion if he judges of service and propriety rather than of show. The Zouave drill doubtless will become engrafted upon our military system, for in *guerrilla* war, or in charges and assaults, the Zouave system of practice is superior to that of the received tactics for infantry. What we inveigh against is, not the system, but the gay colors and cumbrous nature of the costume adopted.

There is really great *art* in the choice of a uniform, and the volunteer should understand it before adopting his regimental costume. Thus, a dark gray dress is preferable, in a professional view, because *it affords less attraction to the enemy's aim.* A Zouave will be detected at a long distance by his red and yellow dress, and will, therefore, make a *good mark* for the enemy, when a gray would so assimilate to the color of the ground as to be unobserved at a distance, or in the night. A scouting-party, or spies, dressed in red breeches would be a commander's folly. The old hunters and Indian fighters understood this *art* of dress, and practiced it in choosing only gray and blue for their outer dress, with leggings of buckskin. They always wore their red flannel next to the skin, and out of sight.

Choose that branch of service—infantry, cavalry, artillery, or naval—to which you are best adapted by taste, by physical strength and by desire to excel.

Make a solemn pledge not to gamble, not to drink ardent spirits, not to swear nor use obscene language if you would preserve your own self-respect as well as the respect of your officers. A soldier's life is embraced by the vicious man from an inclination to indulge his vicious propensities, and it should be the volunteer's aim to elevate the service by frowning down whatever tends to injure and debase the service.

When once enlisted, strive, by all diligence and duty, to attain to perfection in the various exercises of the squad, the company and the regiment. An earnest desire to excel, a close attention to duty, and thoughtful observation will soon render you an expert, to be pointed out by the captain as " one of my best men." A soldier's profession can only be learned by practice and observation. Many a man goes through an entire season's campaign without attaining a knowledge beyond the simplest exercises and maneuvers because of indifference to duty, and inattention.

Discipline has every thing to do with success. Anthony Wayne, with his two Pennsylvania brigades, was considered equal, in combat, to twice his force of the enemy, because his discipline was so rigid. So with all great commanders. No man ever made a great captain who did not control his men with the most mathematical certainty. The soldier will not, therefore, complain at the severity of a superior who exacts the most explicit obedience to all orders, for, in hours of danger, he can rely upon that officer as his leader, and the severe discipline enforced may be the means of bringing a victory, where a lax discipline would surely bring defeat and disgrace.

As very much of the efficiency of the soldier depends upon the state of his health, particular care should be taken to preserve that health. Let the utmost attention be given to habits, to food and drink, to sleeping, to the state of the body in regard to cleanliness, so far as circumstances will allow. Even poor food, well prepared, will conduce to health, whereas good food, poorly prepared, will prove deleterious. We give, herewith, such hints, recipes and general directions as will enable the soldier to pass through hard service without forsaking all comforts, or foregoing all the satisfaction derived from healthy food and sound sleep.

Finally, in your entire demeanor and habits, be exemplary, steady, studious. Observe all the regulations of the army to the letter. Be not remiss in your respect of the Sabbath, and all religious exercises of your chaplain. Remember that it is better to die on the field of battle as a Christian should die, than to die as one careless of his relations to the great Hereafter. With a heart open to generous impulses, be as firm and invincible to duty as steel—as true to your cause as the stars to the mariner.

The soldier, embodying these suggestions, will gain laurels, will command respect and be sure of promotion ; while he who discards all counsels to duty and right will be certain to come forth from the service without honor, without that good name which should be his most blessed inheritance.

GENERAL ORDERS, REGULATIONS, ETC.

Who Can Enlist.

The same rule applies to volunteers that prevails in the regular service, in regard to enlistment, viz. :—" Any free white male person above the age of eighteen and under thirty-five, being at least five feet, four and a half inches high, effective, able-bodied, sober, free from disease, of good character and habits, and with a competent knowledge of the English language, may be enlisted." " No person who is under the age of twenty-one years is to be enlisted without the written consent of his parent, guardian or master." Of course large numbers of volunteers are enlisted who do not meet all these requisitions. Some are discarded before inspection by the State authorities. Others again do not pass that inspection, and are dropped from the roll ; and then, after the company or regiment is ordered into Government service, another inspection is made by the regular army authorities, whereby any defective or incompetent man is liable to be ordered out. Notwithstanding all this routine, however, many a bad man, many drunkards, many unsound men " pass muster." The fact that an entire regiment has recently been enlisted from the professional thieves and pickpockets of New York city shows that the regulations are sometimes compromitted, to suit special cases.

Equipments.

The different States have authorized slightly different equipments for their volunteers, though they all conform, very nearly, to the Federal Government provision for troops in the regular service. New York furnishes to each private volunteer :—one jacket, one pair trowsers, one over-coat, two flannel shirts, two flannel drawers, two pairs woolen socks, one pair shoes, one fatigue cap, one blanket, one knapsack, one haversack, and one canteen. Also, one tin cup, plate, spoon, etc. This is the *preliminary* outfit. When the troops are mustered into Government service, they receive regular Government allowances, according to the terms of their service. Officers provide their own equipments.

Election of Officers.

After enrollment and classification into companies, the election or choice of officers takes place. Captains, subalterns and non-commissioned officers are elected by the votes (written ballots) of their respective companies. Great care should be taken by the men that their choice is carefully made, for *their whole comfort and happiness is in the hands of the captain and his subordinates.*

Field-officers of regiments and separate battalions, brigadier-generals and brigade inspectors are chosen by the field-officers of their respective brigades. Staff-officers are selected by their chief officers, viz.:— major-generals, brigadier-generals and commanding officers of regiments or separate battalions appoint staff-officers to their respective commands.

Transfers.

No non-commissioned officer or soldier can be transferred from one regiment to another without the authority of the commanding general. The colonel may, however, upon the application of the captains, transfer a non-commissioned officer or soldier from one company to another of the same regiment, but then only with the consent of the department commander in case of change of post. When soldiers are authorized to be transferred the same will take place on the first of the month. In all cases of transfer, a complete descriptive roll will accompany the soldier transferred, which roll will embrace an account of his pay, clothing and other allowances; also all stoppages to be made on account of the Government and debts due the washerwomen, as well as such other facts as may be necessary to show his character and military history.

Pay.

See tabular statement of pay to all classes and grades of officers and men, pages 58–59–60–61–62–63. Volunteers receive the same pay as regulars after their acceptance by the State. The State pays through its paymaster-general its troops until they are sworn into the general Government service, when all State responsibility ceases.

Extra work is paid for when soldiers are required to work on fortifications, on artillery roads, making surveys, or any other unusual labor of ten days' duration or more, viz.:— twenty-five cents per day.

Special bounties are paid for enlistments in the regular service, at various stations, thus:—for enlistment at or near any of the posts in Texas, $26 bounty ; in New Mexico, $52 ; in California, $117 ; in Oregon, or in Washington Territory, $142 ; at or near Fort Snelling, $23 ; at or near Great Salt Lake City, $85, etc.   Privates, in the regular army, are also given *certificates of merit* for good conduct, upon which they draw two dollars extra, per month.   Three months *extra pay* is also granted to every non-commissioned officer, musician or private on re-enlistment, after expiration of his first term.

The Oath of Allegiance.

The volunteer is not required to take any special oath upon his enlistment further than the paper which is signed at the time of such enlistment ; but when the troops are mustered into the general Government service the following oath is administered :—

"I, A. B., do solemnly swear (or affirm) that I will bear true allegiance to the United States of America, and that I will serve them honestly and faithfully against all their enemies or opposers whatsoever, and observe and obey the orders of the President of the United States, and the orders of the officers appointed over me according to the rules and articles for the government of the armies of the United States."

Rations.

When once in the service the subsistence rations are as follows, per day, which it is the captain's duty to see is provided, regularly and of good quality :

ARTICLES.	ONE RATION.	100 RATIONS.
Pork,	12 oz.	75 lbs.
Beef (salt) (in lieu of pork),	20 oz.	125 lbs.
Beef (fresh) (in lieu of salt beef),	20 oz.	125 lbs.
Flour,	18 oz.	112 1-2 lbs.
Hard bread (in lieu of flour),	12 oz.	75 lbs.
Beans or peas,	—	8 qts, or 15 lbs.
Coffee,	—	10 lbs.
Sugar,	—	12 lbs.
Vinegar,	—	1 gallon.
Candles,	—	1 1-4 lbs.
Soap,	—	4 lbs.
Salt,	—	2 quarts.

Hard bread *extra*, four ounces per day, per man, at sea, on the march, or in active service. In *lieu of* beans or peas, ten pounds of rice is allowed to every one hundred rations. Watson's edition of Hardee's Tactics adds to each ration— desiccated mixed vegetables, one ounce ; desiccated potatoes, one and one-half ounce ; but by what authority the addition is made we know not.

The army regulations specify that when a soldier is detached on duty, and it is impracticable for him to carry his subsistence with him, it will be commuted at seventy-five cents per day. The ration of a soldier stationed in a city, with no opportunity of messing, will be commuted at forty cents. The rations of soldiers on furlough, or on stations where they can not be issued in kind, are commuted at the cost or value at the post. When a soldier on duty has paid his own subsistence from necessity, he will have refunded the cost of his regular ration. Extra issues are allowed in certain cases, of fresh vegetables, pickled onions, sourkrout, molasses, dried apples, etc., as specified in regulation 1079 of " Army Regulations." The soldier is of course at liberty to add to his daily rations such articles of food as he may purchase. The army is rarely ever in a locality where such purchases may not be made of the sutler or commissary.

Daily Routine Rules.

Once in camp, the soldier enters upon military life in earnest. As he is expected to consult this manual to understand what his daily routine of duty and habit is, we will here give the official regulations at length, as prescribed for companies in camp or barracks :

The captain will cause the men of the company to be numbered, in a regular series, including the non-commissioned officers, and divided into four squads, each to be put under the charge of a non-commissioned officer.

Each subaltern officer will be charged with a squad for the supervision of its order and cleanliness ; and captains will require their lieutenants to assist them in the performance of *all* company duties.

As far as practicable, the men of each squad will be quartered together.

The utmost attention will be paid by commanders of

companies to the cleanliness of their men, as to their persons, clothing, arms, accouterments and equipments, and also as to their quarters or tents.

The name of each soldier will be labeled on his bunk, and his company number will be placed against his arms and accouterments.

The arms will be placed in the arm-racks, the stoppers in the muzzles, the cocks let down, and the bayonets in their scabbards; the accouterments suspended over the arms, and the swords hung up by the belts on pegs.

The knapsack of each man will be placed on the lower shelf of his bunk, at its foot, packed with his effects, and ready to be slung; the great-coat on the same shelf, rolled and strapped; the coat, folded inside out, and placed under the knapsack; the cap on the second or upper shelf; and the boots well cleaned.

Dirty clothes will be kept in an appropriate part of the knapsack; no article of any kind to be put . under the bedding.

Cooking utensils and table equipage will be cleaned and arranged in closets or recesses; blacking and brushes out of view; the fuel in boxes.

Ordinarily the cleaning will be on Saturdays. The chiefs of the squads will cause bunks and bedding to be overhauled; floors dry rubbed; tables and benches scoured; arms cleaned; accouterments whitened and polished, and every thing put in order.

Where conveniences for bathing are to be had, the men should bathe once a week. The feet to be washed at least twice a week. The hair *kept short*, and beard neatly trimmed.

Non-commissioned officers, in command of squads, will be held more immediately responsible that their men observe what is prescribed above; that they wash their hands and faces daily; that they brush or comb their heads; that those who are to go on duty put their arms, accouterments, dress, etc., in the best order, and that such as have permission to pass the chain of sentinels are in the dress that may be ordered.

Commanders of companies and squads will see that the

arms and accouterments in possession of the men are always kept in good order, and that proper care be taken in cleaning them.

When belts are given to a soldier, the captain will see that they are properly fitted to the body ; and it is forbidden to cut any belt without his sanction.

Cartridge-boxes and bayonet-scabbards will be polished with blacking ; varnish is injurious to the leather, and will not be used.

All arms in the hands of the troops, whether browned or bright, will be kept in the state in which they are issued by the Ordnance Department. Arms will not be taken to pieces without permission of a commissioned officer. Bright barrels will be kept clean and free from rust without polishing them ; care should be taken in rubbing not to bruise or bend the barrel. After firing, wash out the bore ; wipe it dry, and then pass a bit of cloth, slightly greased, to the bottom. In these operations, a rod of wood with a loop in one end is to be used instead of the rammer. The barrel, when not in use, will be closed with a stopper. For exercise, each soldier should keep himself provided with a piece of sole leather to fit the cup or countersink of the hammer.

Arms shall not be left loaded in quarters or tents, or when the men are off duty, except by special orders.

Ammunition issued will be inspected frequently. Each man will be made to pay for the rounds expended without orders, or not in the way of duty, or which may be damaged or lost by his neglect.

Ammunition will be frequently exposed to the dry air, or sunned.

Special care shall be taken to ascertain that no ball-cartridges are mixed with the blank cartridges issued to the men.

All knapsacks are to be painted black. Those for the artillery will be marked in the center of the cover with the number of the regiment only, in figures of one inch and a half in length, of the character called full face with yellow paint. Those for the infantry will be marked in the same way, in white paint. Those for the ordnance will be marked with two cannon, crossing ; the cannon to be seven and a half inches in length, in yellow paint, to resemble those on the cap. The knapsack straps will be black.

The knapsacks will also be marked upon the inner side with the letter of the company and the number of the soldier, on such part as may be readily observed at inspections.

Haversacks will be marked upon the flap with the number and name of the regiment, the letter of the company, and number of the soldier, in black letters and figures. And each soldier must, at all times, be provided with a haversack and canteen, and will exhibit them at all inspections. It will be worn on the left side on marches, guard, and when paraded for detached service—the canteen outside the haversack.

The front of the drums will be painted with the arms of the United States, on a blue field for the infantry, and on a red field for the artillery. The letter of the company and number of the regiment, under the arms in a scroll.

Officers at their stations, in camp or in garrison, will always wear their proper uniform.

Soldiers will wear the prescribed uniform in camp or garrison, and will not be permitted to keep in their possession any other clothing. When on fatigue parties, they will wear the proper fatigue dress.

Soldier's Mess.

An important adjunct of convenience and comfort is the arrangement of the company into messes. The matter is the subject of express regulation. We will give the " official " requisitions, viz. :

In camp or barracks, the company officers must visit the kitchen daily and inspect the kettles, and at all times carefully attend to the messing and economy of their respective companies. The commanding officer of the post or regiment will make frequent inspections of the kitchens and messes.

The bread must be thoroughly baked, and not eaten until it is cold. The soup must be boiled at least five hours, and the vegetables always cooked sufficiently to be perfectly soft and digestible.

Messes will be prepared by privates of squads, including private musicians, each taking his tour. The greatest care will be observed in washing and scouring the cooking utensils; those made of brass and copper should be lined with tin.

The messes of prisoners will be sent to them by the cooks.

No persons will be allowed to visit or remain in the kitchens, except such as may come on duty, or be occupied as cooks.

Those detailed for duty in the kitchens will also be required to keep the furniture of the mess-room in order.

On marches and in the field, the only mess furniture of the soldier will be one tin plate, one tin cup, one knife, fork, and spoon, to each man, to be carried by himself on the march.

If a soldier be required to assist his first sergeant in the writing of the company, to excuse him from a tour of millitary duty, the captain will previously obtain the sanction of his own commander, if he have one present; and whether there be a superior present or not, the captain will be responsible that the man so employed does not miss two successive tours of guard-duty by reason of such employment.

Tradesmen may be relieved from ordinary military duty to make, to alter, or to mend soldiers' clothing, etc. Company commanders will fix the rates at which work shall be done, and cause the men, for whose benefit it is done, to pay for it at the next pay-day.

Each company officer, serving with his company, may take from it one soldier as waiter, with his consent and the consent of his captain. No other officer shall take a soldier as a waiter. Every soldier so employed shall be so reported and mustered.

Soldiers taken as officers' waiters shall be acquainted with their military duty, and at all times be completely armed and clothed, and in every respect equipped according to the rules of the service, and have all their necessaries complete and in good order. They are to fall in with their respective companies at all reviews and inspections, and are liable to such drills as the commanding officer shall judge necessary to fit them for service in the ranks.

Non-commissioned officers will, in no case, be permitted to act as waiters; nor are they, or private soldiers, not waiters, to be employed in any menial office, or made to perform any service not military, for the private benefit of any officer or mess of officers.

Washer-Women.

To each company are allowed four women, who each receive the regular ration of a soldier, but are not otherwise paid. Their duties are those of washer-women to the men. The price of the washing is prescribed, and is paid out of the

soldiers' regular monthly pay.   The women are liable to be
discharged or "drummed out of camp" in event of any gross
misconduct, drunkenness or breach of camp etiquette.   Each
woman is required to have a certificate of good character from
head-quarters before she can assume duty within the lines.

Inspections.

Captains inspect their companies every Sunday morning.
No soldier is excused from the inspection except the guard,
the sick and the necessary attendants on the hospital.

Regimental inspection is made on the last day of each month.
It is preceded by a review, as are all general inspections.

Inspection is also made of troops when mustered for
payment.

Inspection implies a thorough examination of the arms,
accouterments and clothing of the soldier, as well as his own
personal condition.

The hospital is thoroughly inspected, by its officers in
charge, every Sunday morning.

Muster always precedes the review.

A daily dress parade is always made.

The army rules which govern musters, reviews, inspections
and parades are only to be learned from the official work,
Article xxx.

Issues of Rations

Depend upon circumstances.   When an army is not moving
rations are generally issued for four days at a time.   Issues to
the companies of a regiment, and the fatigues to receive them,
are superintended by an officer detailed from the regiment.

Straw.

Twelve pounds of straw per month, is allowed in barracks
for each man, servant and company woman, for bedding.   One
hundred pounds per month is allowed each horse.   In camp
it is not used to any extent for bedding, except the camp
becomes barracks, in permanency.   The commissary is, as a
general thing, able to supply it to all who can produce an
order for it from the proper officer.

Watchwords, Countersigns, etc.

See Dictionary of Military Terms comprised in this work,
for the nature and uses of these terms.

Hours of Service and Roll Call.

The duties of the day commence, in garrison, with the morning *reveille*, which sounds at five o'clock in May, June, July and August; at six o'clock in March, April, September and October; and at half-past six in November, December, January and February. The *troop call, surgeon's call, water calls*, breakfast and dinner *signals* are prescribed by the commanding officer according to season, climate and circumstances. In cavalry, *stable calls* immediately after *reveille.*

In camp, the commanding officer prescribes the hours of reveille reports, roll calls, guard mounting, meals, stable calls, issues, fatigues, etc.

Signals.

The several signals are as follows:

1. To go for fuel—*poing stroke and ten stroke roll.*
2. To go for water—*two strokes and a flam.*
3. For fatigue party—*pioneer's march.*
4. Adjutant's call—*first part of the troop.*
5. 1st sergeant's call—*one roll and four taps.*
6. Sergeant's call—*one roll and three taps.*
7. Corporal's call—*one roll and two taps.*
8. For the drummers—*the drummer's call.*

Roll Calls

Are three daily, viz.: *reveille, retreat* and *tattoo*, and one made on company parades by the first sergeants, superintended by a commissioned officer of the company. Captains are instructed to report all absentees without leave to the colonel or commanding officer. Immediately after the reveille roll call (or after stable duty in the cavalry) the tents or quarters, and the space around them, will be put in order by the men of the companies, superintended by the chiefs of the squads, and the guard house or guard tent by the guard or prisoners.

Guards.

The importance of the guard especially commends the subject to the volunteer for his early study. We may here give the official regulations entire as affording the proper preliminary school for the novice:

Sentinels will be relieved every two hours, unless the state of the weather, or other causes, should make it necessary or proper that it be done at shorter or longer intervals.

Each relief, before mounting, is inspected by the commander of the guard or of its post. The corporal reports to him, and presents the old relief on its return.

The *countersign*, or watchword, is given to such persons as are entitled to pass during the night, and to officers, non-commissioned officers, and sentinels of the guard. Interior guards receive the countersign only when ordered by the commander of the troops.

The *parole* is imparted to such officers only as have a right to visit the guards, and to make the grand rounds; and to officers commanding guards.

As soon as the new guard has been marched off, the officer of the day will repair to the office of the commanding officer and report for orders.

The officer of the day must see that the officer of the guard is furnished with the parole and countersign before *retreat.*

The officer of the day visits the guards during the day at such times as he may deem necessary, and makes his rounds at night at least once after twelve o'clock.

Upon being relieved, the officer of the day will make such remarks in the report of the officer of the guard as circumstances require, and present the same at head-quarters.

Commanders of guards leaving their posts to visit their sentinels, or on other duty, are to mention their intention, and the probable time of their absence, to the next in command.

The officers are to remain constantly at their guards, except while visiting their sentinels, or necessarily engaged elsewhere on their proper duty.

Neither officers nor soldiers are to take off their clothing or accouterments while they are on guard.

The officer of the guard must see that the countersign is duly communicated to the sentinels a little before twilight.

When a fire breaks out, or any alarm is raised in a garrison, all guards are to be immediately under arms.

Inexperienced officers are put on guard as supernumeraries, for the purpose of instruction.

Sentinels will not take orders or allow themselves to be relieved, except by an officer or non-commissioned officer of their guard or party, the officer of the day, or the commanding officer; in which case the orders will be immediately notified to the commander of the guard by the officer giving them.

Sentinels will report every breach of orders or regulations they are instructed to enforce.

Sentinels must keep themselves on the alert, observing every thing that takes place within sight and hearing of their post. They will carry their arms habitually at support, or on either shoulder, but will never quit them. In wet weather, if there be no sentry-box, they will secure arms.

No sentinel shall quit his post or hold conversation not necessary to the proper discharge of his duty.

All persons, of whatever rank in the service, are required to observe respect toward sentinels.

In case of disorder, a sentinel must call out *the guard ;* and if a fire take place, he must cry—" *Fire !*" adding the number of his post. If in either case the danger be great, he must discharge his firelock before calling out.

It is the duty of a sentinel to repeat all calls made from posts more distant from the main body of the guard than his own, and no sentinel will be posted so distant as not to be heard by the guard, either directly or through other sentinels.

Sentinels will present arms to general and field officers, to the officer of the day, and to the commanding officer of the post. To all other officers they will carry arms.

When a sentinel in his sentry-box sees an officer approaching, he will stand at *attention*, and as the officer passes will salute him, by bringing the left hand briskly to the musket, as high as the right shoulder.

The sentinel at any post of the guard, when he sees any body of troops, or an officer entitled to compliment, approach, must call—" *Turn out the guard !*" and announce who approaches.

Guards do not turn out as a matter of compliment after sunset ; but sentinels will, when officers in uniform approach, pay them proper attention, by facing to the proper front, and standing steady at *shouldered arms*. This will be observed until the evening is so far advanced that the sentinels begin challenging.

After retreat (or the hour appointed by the commanding officer), until broad daylight, a sentinel challenges every person who approaches him, taking, at the same time, the position of *arms port.* He will suffer no person to come nearer than within reach of his bayonet, until the person has given the countersign.

A sentinel, in challenging, will call out—" *Who comes there ?*" If answered—" *Friend, with the countersign,*" and he be instructed to pass persons with the countersign, he will reply—" *Advance, friend, with the countersign !*" If answered —" *Friends !*" he will reply—" *Halt, friends ! Advance one with the countersign !*" If answered—" *Relief,*" " *Patrol,*" or " *Grand rounds,*" he will reply—" *Halt ! Advance, sergeant (or corporal), with the countersign !*" and satisfy himself that the party is what it represents itself to be. If he have no authority to pass persons with the countersign, if the wrong countersign be given, or if the persons have not the countersign, he will cause them to stand, and call—" *Corporal of the guard !*"

In the daytime, when the sentinel before the guard sees the officer of the day approach, he will call—" *Turn out the guard ! officer of the day.*" The guard will be paraded, and salute with presented arms.

When any person approaches a post of the guard at night, the sentinel before the post, after challenging, causes him to halt until examined by a non-commissioned officer of the guard. If it be the officer of the day, or any other officer entitled to inspect the guard and to make the rounds, the non-commissioned officer will call—" *Turn out the guard !*" when the guard will be paraded at shouldered arms, and the officer of the guard, if he thinks necessary, may demand the countersign and parole.

The officer of the day, wishing to make the rounds, will take an escort of a non-commissioned officer and two men. When the rounds are challenged by a sentinel, the sergeant will answer—" *Grand rounds !*" and the sentinel will reply— " *Halt, grand rounds ! Advance, sergeant, with the countersign !*" Upon which the sergeant advances and gives the countersign. The sentinel will then cry—" *Advance, rounds !*" and stand at a shoulder till they have passed.

When the sentinel before the guard challenges, and is answered—" *Grand rounds,*" he will reply—" *Halt, grand rounds! Turn out the guard; grand rounds!"* Upon which the guard will be drawn up at shouldered arms. The officer commanding the guard will then order a sergeant and two men to advance; when within ten paces, the sergeant challenges. The sergeant of the grand rounds answers—" *Grand rounds!"* The sergeant of the guard replies—"*Advance, sergeant, with the countersign!"* The sergeant of the rounds advances alone, gives the countersign, and returns to his round. The sergeant of the guard calls to his officer—" *The countersign is right!"* on which the officer of the guard calls— "*Advance, rounds!"* The officer of the rounds then advances alone, the guard standing at shouldered arms. The officer of the rounds passes along the front of the guard to the officer, who keeps his post on the right, and gives him the parole. He then examines the guard, orders back his escort, and, taking a new one, proceeds in the same manner to other guards.

All material instructions given to a sentinel on post by persons entitled to make grand rounds, ought to be promptly notified to the commander of the guard.

Any general officer, or the commander of a post or garrison, may visit the guards of his command, and go the grand rounds, and be received in the same manner as prescribed for the officer of the day.

The Police Guard.

In each regiment a police guard is detailed every day, consisting of two sergeants, three corporals, two drummers, and men enough to furnish the required sentinels and patrols. The men are taken from all the companies, from each in proportion to its strength. The guard is commanded by a lieutenant, under the supervision of a captain, as regimental officer of the day. It furnishes ten sentinels at the camp: one over the arms of the guard; one at the colonel's tent; three on the color front, one of them over the colors; three fifty paces in rear of the field officers' tents; and one on each flank, between it and the next regiment. If it is a flank regiment, one more sentinel is posted on the outer flank.

An advanced post is detached from the police guard, composed of a sergeant, a corporal, a drummer and nine men to furnish sentinels and the guard over the prisoners. The men are the first of the guard roster from each company. The men of the advanced post must not leave it under any pretext. Their meals are sent to the post. The advanced post furnishes three sentinels ; two a few paces in front of the post, opposite the right and left wing of the regiment, posted so as to see as far as possible to the front, and one over the arms.

In the cavalry, dismounted men are employed in preference on the police guard. The mounted men on guard are sent in succession, a part at a time, to groom their horses. The advanced post is always formed of mounted men.

In each company, a corporal has charge of the stable-guard. His tour begins at retreat, and ends at morning stable-call. The stable-guard is large enough to relieve the men on post every two hours. They sleep in their tents, and are called by the corporal when wanted. At retreat he closes the streets of the camp with cords, or uses other precautions to prevent the escape of loose horses.

The officer of the day is charged with the order and cleanliness of the camp ; a fatigue is furnished to him when the number of prisoners is insufficient to clean the camp. He has the calls beaten by the drummer of the guard.

The police guard and the advanced post pay the same honors as other guards. They take arms when an armed body approaches.

The sentinel over the colors has orders not to permit them to be moved except in presence of an escort ; to let no one touch them but the color-bearer, or the sergeant of the police guard when he is accompanied by two armed men.

The sentinels on the color front permit no soldier to take arms from the stacks, except by order of some officer, or a non-commissioned officer of the guard. The sentinel at the colonel's tent has orders to warn him, day or night, of any unusual movement in or about the camp.

The sentinels on the front, flanks, and rear, see that no soldier leaves camp with horse or arms unless conducted by a non-commissioned officer. They prevent non-commissioned officers and soldiers from passing out at night, except to go to the sinks, and mark if they return. They arrest, at any time, suspicious persons prowling about the camp, and at night, every one who attempts to enter, even the soldiers of other corps. Arrested persons are sent to the officer of the guard, who sends them, if necessary, to the officer of the day.

The sentinels on the front of the advanced post have orders to permit neither non-commissioned officers nor soldiers to pass the line, without reporting at the advanced post; to warn the advanced post of the approach of any armed body, and to arrest all suspicious persons. The sergeant sends persons so arrested to the officer of the guard, and warns him of the approach of any armed body.

The sentinel over the arms at the advanced post guards the prisoners and keeps sight of them, and suffers no one to converse with them without permission. They are only permitted to go to the sinks one at a time, and under a sentinel.

If any one is to be passed out of camp at night, the officer of the guard sends him under escort to the advanced post, and the sergeant of the post has him passed over the chain.

At retreat, the officer of the guard has the roll of his guard called, and inspects arms, to see that they are loaded and in order; and visits the advanced post for the same purpose. The sergeant of the police guard, accompanied by two armed soldiers, folds the colors and lays them on the trestle in rear of the arms. He sees that the sutler's stores are then closed, and the men leave them, and that the kitchen fires are put out at the appointed hour.

The officer of the day satisfies himself frequently during the night, of the vigilance of the police guard and advanced post. He prescribes patrols and rounds to be made by the officer and non-commissioned officers of the guard. The officer of the guard orders them when he thinks necessary. He visits the sentinels frequently.

At reveille, the police guard takes arms; the officer of the guard inspects it and the advanced post. The sergeant replants the colors in place. At retreat and reveille the advanced post takes arms; the sergeant makes his report to the officer of the guard when he visits the post.

When necessary, the camp is covered at night with small outposts, forming a double chain of sentinels. These posts are under the orders of the commander of the police guard, and are visited by his patrols and rounds.

The officer of the guard makes his report of his tour of
service, including the advanced post, and sends it, after the
guard is marched off, to the officer of the day.

When the regiment marches, the men of the police guard
return to their companies, except those of the advanced post.
In the cavalry, at the sound, " boot and saddle," the officer
of the guard sends one-half the men to saddle and pack;
when the regiment assembles, all the men join it.

When the camping-party precedes the regiment, and the
new police guard marches with the camping-party, the guard,
on reaching the camp, forms in line thirty paces in front of
the center of the ground marked for the regiment. The
officer of the guard furnishes the sentinels required by the
commander of the camping-party. The advanced post takes
its station.

The advanced post of the old police guard takes charge of
the prisoners on the march, and marches, bayonets fixed, at
the center of the regiment. On reaching camp, it turns over
the prisoners to the new advanced post.

The picket guard and grand guard are so involved in
general orders as to render the detail of their performance
unnecessary here. They concern officers more than the men,
and their service is to be learned from the commanding
officer.

On the March.

Thus far we have given an insight into camp or stationary
duty and life—a life the true soldier soon learns to grow
weary of, and to be " on the march" is his daily wish. We
will then show him what that really is, as applied to the
movement of an entire division or army.

The object of the movement and the nature of the ground
determine the order of march, the kind of troops in each
column, and the number of columns.

The force is divided into as many columns as circumstances
permit, without weakening any one too much. They ought
to preserve their communications, and be within supporting
distance of each other. The commander of each column
ought to know the strength and direction of the others.

The advance and rear guards are usually light troops; their
strength and composition depend on the nature of the ground

and the position of the enemy. They serve to cover the movements of the army, and to hold the enemy in check until the general has time to make his arrangements.

The advance guard is not always at the head of the column; in a march to a flank, it takes such positions as cover the movement. Sappers are attached to the advanced guard if required.

The "*general*," sounded one hour before the time of marching, is the signal to strike tents, to load the wagons, and pack horses, and send them to the place of assembling. The fires are then put out, and care taken to avoid burning straw, etc., or giving to the enemy any other indication of the movement.

The "march" will be beat in the infantry, and the "advance" sounded in the cavalry, in succession, as each is to take its place in the column.

When the army should form suddenly to meet the enemy, the "*long roll*" is beat, and "*to horse*" sounded. The troops form rapidly in front of their camp.

Batteries of artillery and their caissons move with the corps to which they are attached; the field train and ambulances march at the rear of the column; and the baggage with the rear-guard.

Cavalry and infantry do not march together, unless the proximity of the enemy makes it necessary.

In cavalry marches, when distant from the enemy, each regiment, and, if possible, each squadron, forms a separate column, in order to keep up the same gait from front to rear, and to trot, when desirable, on good ground. In such cases, the cavalry may leave camp later, and can give more rest to the horses, and more attention to the shoeing and harness. Horses are not bridled until the time to start.

When necessary, the orders specify the rations the men are to carry in their haversacks. The field officers and captains make inspections frequently during the march; at halts they examine the knapsacks, valises, and haversacks, and throw away all articles not authorized. The officers and non-commissioned officers of cavalry companies attend personally to the packs and girths.

When it can be avoided, troops should not be assembled

on high roads or other places where they interrupt the communication.

Generals of division and commanders of detached corps send a staff officer to the rendezvous, in advance, to receive the troops, who, on arriving, take their place in the order of battle, and form in close column, unless otherwise ordered. Artillery, or trains halted on the roads, form in file on one side.

The execution of marching orders must not be delayed. If the commander is not at the head of his troops when they are to march, the next in rank puts the column in motion.

If possible, each column is preceded by a detachment of sappers, to remove obstacles to the march, aided, when necessary, by infantry, or the people of the country. The detachment is divided into two sections: one stops to remove the first obstacle, the other moves on to the next.

In night marches, and at bad places, and at cross-roads, when necessary, intelligent non-commissioned officers are posted to show the way, and are relieved by the regiments as they come up.

On the march no one shall fire a gun, or cry "*halt*" or "*march*" without orders.

Soldiers are not to stop for water; the canteens should be filled before starting.

It is better to avoid villages; but if the route lies through them, officers and non-commissioned officers are to be vigilant to prevent straggling. Halts should not take place at villages.

Besides the rear-guard, the general sometimes takes a detachment from the last regiment, and adds to it non-commissioned officers from each regiment, to examine villages and all hiding-places on the route, to bring up stragglers and seize marauders.

In night marches, the sergeant-major of each regiment remains at the rear with a drummer, to give notice when darkness or difficulty stops the march. In cavalry, a trumpet is placed in rear of each squadron, and the signal repeated to the head of the regiment.

The general and field officers frequently stop, or send officers to the rear, to see that the troops march in the pre-scribed order, and keep their distances. To quicken the march, the general warns the colonels, and may order a signal to be beat. It is repeated in all the regiments.

In approaching a defile the colonels are warned; they close their regiments as they come up; each regiment passes separately, at an accelerated pace, and in as close order as possible. The leading regiment having passed, and left room enough for the whole column in close order, then halts, and moves again as soon as the last regiment is through. In the cavalry, each squadron, before quickening the pace to rejoin the column, takes its original order of march.

When the distance from the enemy permits, each regiment, after closing up in front and rear of the defile, stacks arms.

Halts to rest and re-form the troops are frequent during the day, depending on the object and length of the march. They are made in preference after the passage of defiles.

No honors are paid by the troops on the march or at halts.

The sick march with the wagons.

Led horses of officers, and the horses of dismounted men, follow their regiment. The baggage wagons never march in the column. When the general orders the field train and ambulances to take place in the column, he designates the position they shall take.

If two corps meet on the same road, they pass to the right, and both continue their march, if the road is wide enough; if it is not, the first in the order of battle takes the road, the other halts.

A corps in march must not be cut by another. If two corps meet at cross-roads, that which arrives last halts if the other is in motion. A corps in march passes a corps at a halt, if it has precedence in the order of battle, or if the halted corps is not ready to move at once.

A column that halts to let another column pass resumes the march in advance of the train of this column. If a column has to pass a train, the train must halt, if necessary, till the column passes. The column which has precedence must yield it if the commander, on seeing the orders of the other, finds it for the interest of the service.

Battle.

We now come to the great event, the conflict—pregnant

with such momentous results to all concerned, bringing death, wounds, pain, victory, defeat, all in its sad, fearful train of consequences. It is well to contemplate the manner in which the great tragedy is to be enacted. We, hence, give the official announcements of the order of the day, so far as it can be regulated by the authorities at Washington.

Dispositions for battle depend on the number, kind, and quality of the troops opposed, on the ground, and on the objects of the war; but the following rules are to be observed generally:

In attacking, the advance guard endeavors to capture the enemy's outposts, or cut them off from the main body. Having done so, or driven them in, it occupies, in advancing, all the points that can cover or facilitate the march of the army, or secure its retreat, such as bridges, defiles, woods, and heights; it then makes attacks, to occupy the enemy, without risking too much, and to deceive them as to the march and projects of the army.

When the enemy is hidden by a curtain of advanced troops, the commandant of the advanced guard sends scouts, under intelligent officers, to the right and left, to ascertain his position and movements. If he does not succeed in this way, he tries to unmask the enemy by demonstrations; threatens to cut the advance from the main body; makes false attacks; partial and impetuous charges in echelon; and if all fail, he makes a real attack to accomplish the object.

Detachments left by the advanced guard to hold points in the rear rejoin it when other troops come up. If the army takes a position, and the advanced guard is separated from it by defiles or heights, the communication is secured by troops drawn from the main body.

At proper distance from the enemy, the troops are formed for the attack in several lines; if only two can be formed, some battalions in column are placed behind the wings of the second line. The lines may be formed of troops in column or in order of battle, according to the ground and plan of attack.

The advanced guard may be put in the line or on the wings, or other position, to aid the pursuit or cover the retreat.

The reserve is formed of the best troops of foot and horse,

to complete a victory or make good a retreat. It is placed in the rear of the center, or chief point of attack or defense.

The cavalry should be distributed in echelon on the wings, and at the center on favorable ground.

It should be instructed not to take the gallop until within charging distance; never to receive a charge at a halt, but to meet it, or, if not strong enough, to retire maneuvering; and in order to be ready for the pursuit, and prepared against a reverse, or the attacks of the reserve, not to engage all its squadrons at once, but to reserve one-third, in column or in echelon, abreast of or in the rear of one of the wings; this arrangement is better than a second line with intervals.

In the attack, the artillery is employed to silence the batteries that protect the position. In the defense it is better to direct its fire on the advancing troops. In either case, as many pieces are united as possible, the fire of artillery being formidable in proportion to its concentration.

In battles and military operations it is better to assume the offensive, and put the enemy on the defensive; but to be safe in doing so requires a larger force than the enemy, or better troops, and favorable ground. When obliged to act on the defensive, the advantage of position and of making the attack may sometimes be secured by forming in rear of the ground on which we are to fight, and advancing at the moment of action. In mountain warfare, the assailant has always the disadvantage; and even in offensive warfare, in the open field, it may frequently be very important, when the artillery is well posted, and any advantage of ground may be secured, to await the enemy and compel him to attack.

The attack should be made with a superior force on the decisive point of the enemy's position, by masking this by false attacks and demonstrations on other points, and by concealing the troops intended for it by the ground, or by other troops in their front.

Besides the arrangements which depend on the supposed plan of the enemy, the wings must be protected by the ground, or supported by troops in echelon; if the attack of the enemy is repulsed, the offensive must at once be taken, to inspire the troops to disconcert the enemy, and often to decide the action. In thus taking the offensive, a close column should be pushed rapidly on the wing or flank of the enemy. The divisions of this column form in line of battle successively, and each division moves to the front as soon as formed, in order, by a rapid attack in echelon, to prevent the enemy from changing front or bringing up his reserves. In all arrangements, especially in those for attack, it is most important to conceal the design until the moment of execution, and then to execute it with the greatest rapidity. The night, therefore, is preferred for the movement of troops on the flank or rear of the enemy, otherwise it is necessary to mask their march by a grand movement in front, or by taking a wide circuit.

In making an attack, the communication to the rear and for retreat must be secured, and the general must give beforehand all necessary orders to provide for that event.

When a success is gained, the light troops should pursue the enemy promptly and rapidly. The other troops will restore order in their columns, then advance from position to position, always prepared for an attack or to support the troops engaged.

Before the action, the generals indicate the places where they will be; if they change position, they give notice of it, or leave a staff officer to show where they have gone.

During the fight the officers and non-commissioned officers keep the men in the ranks, and enforce obedience if necessary. Soldiers must not be permitted to leave the ranks to strip or rob the dead, nor to assist the wounded, unless by express permission, which is only to be given after the action is decided. The highest interest and duty is to win the victory, which only can insure proper care of the wounded.

Before the action, the quartermaster of the division makes all the necessary arrangements for the transportation of the wounded. He establishes the ambulance depôts in the rear, and gives his assistants the necessary instruction for the service of the ambulance wagons and other means of removing the wounded.

The ambulance depôt, to which the wounded are carried or directed for immediate treatment, is generally established at the most convenient building nearest the field of battle. A *red flag* marks its place, or the way to it, to the conductors of the ambulances and to the wounded who can walk.

The active ambulance follow the troops engaged to succor the wounded and remove them to the depôts ; for this purpose the conductors should always have the necessary assistants, that the soldiers may have no excuse to leave the ranks for that object.

The medical director of the division, after consultation with the quartermaster-general, distributes the medical officers and hospital attendants at his disposal, to the depôts and active ambulances. He will send officers and attendants, when practicable, to the active ambulances, to relieve the wounded who require treatment before being removed from the ground. He will see that the depôts and ambulances are provided with the necessary apparatus, medicines, and stores. He will take post and render his professional services at the principal depôt.

If the enemy endanger the depôt, the quartermaster takes the orders of the general to remove it or to strengthen its guard.

The wounded in the depôts and the sick are removed, as soon as possible, to the hospitals that have been established by the quartermaster-general of the army on the flanks or rear of the army.

After an action, the officers of ordnance collect the munitions of war left on the field, and make a return of them to the general. The quartermaster's department collects the rest of the public property captured, and makes the returns to head-quarters.

Written reports for the general commanding-in-chief are made by commandants of regiments, batteries, and separate squadrons, and by all commanders of a higher grade, each in what concerns his own command, and to his immediate commander.

When an officer or soldier deserves mention for conduct in action, a special report shall be made in his case, and the general commanding-in-chief decides whether to mention him in his report to the Government and in his orders. But he shall not be mentioned in the report until he has been mentioned in the orders to the army. These special reports are examined with care by the intermediate commanders, to verify the facts, and secure commendation and rewards to the meritorious only.

The report of battles, which must frequently be made before

these special reports of persons are scrutinized, is confined to general praise or blame, and an account of the operations.

Courtesies.

The law in regard to courtesies to superiors is very strict, and is enforced in both army and navy to its fullest extent. The following are the rules for observance :

The *President* or *Vice-President* is to be saluted with the highest honors—all standards and colors dropping, officers and troops saluting, drums beating and trumpets sounding.

A *general commanding-in-chief* is to be received—by cavalry, with sabers presented, trumpets sounding the march, and all the officers saluting, standards dropping; by infantry, with drums beating the march, colors dropping, officers saluting, and arms presented.

A *major-general* is to be received—by cavalry, with sabers presented, trumpets sounding twice the trumpet-flourish, and officers saluting; by infantry, with three ruffles, colors dropping, officers saluting, and arms presented.

A *brigadier-general* is to be received—by cavalry, with sabers presented, trumpets sounding once the trumpet-flourish, and officers saluting; by infantry, with two ruffles, colors dropping, officers saluting, and arms presented.

An *adjutant-general* or *inspector-general*, if under the rank of a general officer, is to be received at a review or inspection of the troops under arms—by cavalry, with sabers presented, officers saluting; by infantry, officers saluting and arms presented. The same honors to be paid to any field-officer authorized to review and inspect the troops. When the inspecting officer is junior to the officer commanding the parade, no compliments will be paid: he will be received only with swords drawn and arms shouldered.

All guards are to turn out and present arms to *general officers* as often as they pass them, except the personal guards of general officers, which turn out only to the generals whose guards they are, and to officers of superior rank.

To commanders of regiments, garrison, or camp, their own guard turn out, and present arms once a day ; after which, they turn out with shouldered arms.

To the members of the Cabinet ; to the Chief Justice, the President of the Senate, and Speaker of the House of Representa-

tives of the United States; and to Governors, within their respective States and Territories—the same honors will be paid as to a general commanding-in-chief.

Officers of a foreign service may be complimented with the honors due to their rank.

American and foreign envoys or ministers will be received with the compliments due to a major-general.

The colors of a regiment passing a guard are to be saluted, the trumpets sounding, and the drums beating a march.

When general officers, or persons entitled to a salute, pass in the rear of a guard, the officer is only to make his men stand shouldered, and not to face his guard about, or beat his drum.

When general officers, or persons entitled to a salute, pass guards while in the act of relieving, both guards are to salute, receiving the word of command from the senior officer of the whole.

All guards are to be under arms when armed parties approach their posts; and to parties commanded by commissioned officers, they are to present their arms, drums beating a march, and officers saluting.

No compliments by guards or sentinels will be paid between *retreat* and *reveille*, except as prescribed for *grand rounds*.

All guards and sentinels are to pay the same compliments to the officers of the navy, marines, and militia, in the service of the United States, as are directed to be paid to the officers of the army, according to their relative ranks.

It is equally the duty of non-commissioned officers and soldiers, *at all times* and *in all situations*, to pay the proper compliments to officers of the navy and marines, and to officers of other regiments, when in uniform, as to officers of their own particular regiments and corps.

Courtesy among military men is indispensable to discipline. Respect to superiors will not be confined to obedience on duty, but will be extended to all occasions. It is always the duty of the inferior to accost or to offer first the customary salutation, and of the superior to return such complimentary notice.

Sergeants, with swords drawn, will salute by bringing them to a present—with muskets, by bringing the left hand across the body, so as to strike the musket near the right shoulder. Corporals out of the ranks, and privates not sentries, will carry their muskets at a shoulder as sergeants, and salute in like manner.

When a soldier without arms, or with side-arms only, meets an officer, he is to raise his hand to the right side of the visor of his cap, palm to the front, elbow raised as high as the shoulder, looking at the same time in a respectful and soldier-like manner at the officer, who will return the compliment thus offered.

A non-commissioned officer or soldier being seated, and without particular occupation, will rise on the approach of an officer and make the customary salutation. If standing, he will turn toward the officer for the same purpose. If the parties remain in the same place or on the same ground such compliments need not be repeated.

Furloughs.

Furloughs will be granted only by the commanding officer of the post, or the commanding officer of the regiment actually quartered with it. Furloughs may be prohibited at the discretion of the officer in command.

Soldiers on furlough shall not take with them their arms or accouterments.

The form of the furlough is as follows:

TO ALL WHOM IT MAY CONCERN.

The bearer hereof, ——— ———, a sergeant (corporal, or private, as the case may be) of Captain ——— ——— company, ——— regiment of ———, aged ——— years, ——— feet ——— inches high, ——— complexion, ——— eyes, ——— hair, and by profession a ———; born in the ——— of ———, and enlisted at ———, in the ——— of ———, on the ——— day of ———, eighteen hundred and ———, to serve for the period of ———, is hereby permitted to go to ———, in the county of ———, State of ———, he having received a furlough from the ——— day of ———, to the ——— day of ———, at which period he will rejoin his company or regiment at ———, or wherever it then may be, or be considered a deserter.

Subsistence has been furnished to said ——— ——— to the ——— day of ———, and pay to the ——— day of ———, both inclusive.

Given under my hand, at ———, this ——— day of ———, 18———

Signature of the officer ⎱
giving the furlough. ⎰   ——— ———.

Deserters.

We hope we are not writing for any deserter's information. It is not agreeable to give the law to rogues, at any time, but a deserter is one of the basest of rogues; he may betray his brothers' secrets and send them to death or defeat, and we can not think of such a man without disgust. But the soldier must know the law, even in extreme cases, that he may know to what those are amenable who are base enough to dare the ignominy of a deserter's fate. The laws are :

If a soldier desert from, or a deserter be received at, any post other than the station of the company or detachment to which he belonged, he shall be promptly reported by the commanding officer of such post to the commander of his company or detachment. The time of desertion, apprehension, and delivery will be stated. If the man be a recruit, unattached, the required report will be made to the adjutant-general. When a report is received of the apprehension or surrender of a deserter at any post other than the station of the company or detachment to which he belonged, the commander of such company or detachment shall immediately forward his description and account of clothing to the officer making the report.

A reward of thirty dollars will be paid for the apprehension and delivery of a deserter to an officer of the army at the most convenient post or recruiting station. Rewards thus paid will be promptly reported by the disbursing officer to the officer commanding the company in which the deserter is mustered, and to the authority competent to order his trial. The reward of thirty dollars will include the remuneration for all expenses incurred for apprehending, securing, and delivering a deserter.

When non-commissioned officers or soldiers are sent in pursuit of a deserter, the expenses necessarily incurred will be paid whether he be apprehended or not, and reported as in case of rewards paid.

Deserters shall make good the time lost by desertion, unless discharged by competent authority.

No deserter shall be restored to duty without trial, except by the authority competent to order the trial.

Rewards and expenses paid for apprehending a deserter

will be set against his pay, when adjudged by a court-martial, or when he is restored to duty without trial on such condition.

In reckoning the time of service, and the pay and allowances of a deserter, he is to be considered in service when delivered up as a deserter to the proper authority.

An apprehended deserter, or one who surrenders himself, shall receive no pay while waiting trial, and only such clothing as may be actually necessary for him.

Discharges.

No enlisted man shall be discharged before the expiration of his term of enlistment without authority of the War Department, except by sentence of a general court-martial, or by the commander of the department or of an army in the field, on certificate of disability, or on application of the soldier after twenty years' service.

When an enlisted man is to be discharged, his company commander shall furnish him certificates of his account, according to form.

Blank discharges on parchment will be furnished from the adjutant-general's office. No discharge shall be made in duplicate, nor any certificate given in lieu of a discharge.

The cause of discharge will be stated in the body of the discharge, and the space at foot for character cut off, unless a recommendation is given.

Whenever a non-commissioned officer or soldier shall be unfit for the military service in consequence of wounds, disease, or infirmity, his captain shall forward to the commander of the department or of the army in the field, through the commander of the regiment or post, a statement of his case, with a certificate of his disability signed by the senior surgeon of the hospital, regiment or post, according to the form prescribed in the Medical Regulations.

If the recommendation for the discharge of the invalid be approved, the authority therefor will be indorsed on the " certificate of disability," which will be sent back to be completed and signed by the commanding officer, who will then send the same to the adjutant-general's office.

Insane soldiers will not be discharged, but sent, under proper protection, by the department commander to Washington for the order of the War Department for their admission into the Government asylum. The history of the cases, with the men's descriptive list, and accounts of pay and clothing, will be sent with them.

Relative Rank.

The relative rank and command of commissioned and non-commissioned officers in the United States army is as follows, viz. :

1. Lieutenant-general.
2. Major-general.
3. Brigadier-general.
4. Colonel.
5. Lieutenant-colonel.
6. Major.
7. Captain.
8. First lieutenant.
9. Second lieutenant.
10. Cadet.
11. Sergeant-major.
12. Quartermaster sergeant of a regiment.
13. Ordnance sergeant and hospital steward.
14. First sergeant.
15. Sergeant.
16. Corporal.

And in each grade by date of commission or appointment.

Officers serving by *commission* from any State in the Union take rank next *after* officers of the like grade *by commission* from the United States. This rule prescribes the relative rank and command of volunteer or militia commissioned officers in service of the general Government.

When commissions are of the same date the rank is to be decided, between officers of the same regiment or corps, by the *order of appointment ;* between officers of different regiments or corps : 1st, by rank in actual service *when appointed ;* 2d, by former rank and service in the army or marine corps ; 3d, by lottery among such as have not been in the military service of the United States. In case of *equality of rank* by virtue of a brevet commission, reference is had to commissions not brevet. See also 61st and 62d Articles of War.

An officer not having orders from competent authority can not put himself *on duty* by virtue of his commission alone.

Pay Department.

Herewith we give an important table showing the pay, rations, horses and servants given and allowed to every branch of the service :

TABLE OF PAY, SUBSISTENCE, FORAGE, ETC., OF THE U. S. ARMY.

GRADE.	Pay. Per month.	Subsistence. No. of rations per day.	Forage. No. of horses allowed in time of war.	Forage. No. of horses allowed in time of peace.	No. of servants allowed.
Lieutenant-General,	$270 00	40	7	3	4
Major-General,	220 00	15	7	3	4
Senior Aid-de-Camp to General-in-Chief,	80 00		4	3	2
Aid-de-Camp, in addition to pay, etc., of Lieutenant,	24 00		2	1	
Brigadier-General,	124 00	12	5	3	3
Aid-de-Camp to Brigadier, in addition to pay, etc., of Lieutenant,*	20 00		2	1	
Adjutant-General,	110 00	6	5	3	2
Assistant Adjutant-General, with the rank of Lieutenant-Colonel,	95 00	5	4	3	2
Assistant Adjutant-General, with the rank of Major,	80 00	4	4	3	2
Assistant Adjutant-General, with the rank of Captain,	70 00	4	3	1	1
Judge Advocate,	80 00	4	4	3	2
Inspector-General,	110 00	6	5	3	2
Quartermaster-General,	124 00	12	5	3	3
Assistant Quartermaster-General,	110 00	6	5	3	2
Deputy Quartermaster-General,	95 00	5	4	3	2
Quartermaster,	80 00	4	4	3	2

Title					$
Assistant Quartermaster,	1	1	3	4	70 00
Paymaster-General, $2740 per annum.					
Deputy Paymaster-General,	2	3	4	5	95 00
Paymaster,	2	3	4	4	80 00
Commissary-General of Subsistence,	2	3	5	6·	110 00
Assistant Commissary-General of Subsistence,	2	3	4	5	95 00
Commissary of Subsistence, with the rank of Major,	2	3	4	4	80 00
Commissary of Subsistence, with the rank of Captain,	1	1	3	4	70 00
Assistant Commissary of Subsistence, in addition to pay, etc., of Lieutenant,*	·	·	·	·	20 00
Surgeon-General, $2740 per annum.					
Surgeon of ten years' service in that grade,	2	3	4	8	80 00
Surgeon, less than ten years' service,	2	3	4	4	80 00
Assistant Surgeon of ten years' service,	1	1	3	8	70 00
Assistant Surgeon of five years' service,	1	1	3	4	70 00
Assistant Surgeon, less than five years' service,	1	1	2	4	53 33⅓
Superintendent of the Military Academy, not less than the Professor of Natural and Experimental Philosophy. The Commander of Corps of Cadets not less than the Professor of Mathematics.					
Professor of Natural and Experimental Philosophy, $2240 per annum.	1	1	3	4	70 00
Assistant Professor of Natural and Experimental Philosophy,					
Professor of Mathematics, $2240 per annum.	1	1	3	4	70 00
Assistant Professor of Mathematics,					
Professor of Engineering, $2240 per annum.	1	1	3	4	70 00
Ass't Professor of Engineering, and Instructor of Practical Engineering, each,					
Professor of Chemistry, Mineralogy and Geology, $2240 per annum.	1	1	3	4	70 00
Assistant Professor of Chemistry, Mineralogy and Geology, and Assistant Professor of Ethics, each,					
Professor of Ethics, each,					
Chaplain and Professor of Ethics, $2240 per annum.					

TABLE OF PAY, SUBSISTENCE, FORAGE, ETC., OF THE U. S. ARMY.—*Continued.*

GRADE.	Pay. Per month.	Subsistence. No. of rations per day.	Forage. No. of horses allowed in time of war.	No. of horses allowed in time of peace.	No. of servants allowed.
Professor of French, and Professor of Drawing, each $2240 per annum.					
Assistant Professor of French, and Assistant Professor of Drawing, each,	$70 00				
Professor of Spanish, $2240 per annum.					
Instructor of Cavalry and Artillery Tactics,	70 00	4	3	1	1
Adjutant of the Military Academy,	63 33⅓	4	3	1	1
Master of the Sword, $1500 per annum.		4	2	2	1
Teacher of Music,	60 00				
Military Storekeeper, Clothing Department, $1490 per annum.					
Storekeeper of Ordnance at Arsenals of Construction, and in Oregon, California, and New Mexico, $1490 per annum.					
Storekeeper of Ordnance, $1040 per annum.					
Chaplain, to be determined by the Council of Administration, not to exceed,	60 00	4			
Colonel of Engineers, Topographical Engineers, Ordnance, Dragoons, Cavalry, or Mounted Riflemen,	110 00	6	5	3	2
Lieutenant-Colonel of ditto,	95 00	5	4	3	2
Major of ditto,	80 00	4	4	3	2

				Pay	
		2 & 1†	1	70 00	Captain of ditto,
		2 & 1†	1	53 33½	Lieutenant (1st and 2d) of ditto,
				10 00	Adjutant of dragoons, cavalry, or mounted riflemen, in addition to pay, etc., of Lieutenant,
				10 00	Regimental-Quartermaster of ditto,
				21 00	Sergeant-Major of dragoons, cavalry, or mounted riflemen,
				21 00	Quartermaster-Sergeant of do.
				21 00	Chief Bugler of do.
				20 00	First Sergeant of do.
				17 00	Sergeant of do.
				14 00	Corporal of do.
				13 00	Bugler of do.
				15 00	Farrier and Blacksmith of do.
4	3			12 00	Private of do.
				30 00	Master Armorer, Master Carriage-maker, or Master Blacksmith of Ordnance,
				16 00	Armorer, Carriage-maker, or Blacksmith of Ordnance,
				13 00	Artificer of do.
				9 00	Laborer of do.
				22 00	Hospital Steward, appointed by the Secretary of War, and Hospital Steward at posts of more than four companies, pay of Ordnance-Sergeant,
				20 00	Hospital Steward,
				6 00	Matron,

ARTILLERY AND INFANTRY.

				Pay	
6	4	3	2	95 00	Colonel,
5	3	3	2	80 00	Lieutenant-Colonel,
4	3	3	2	70 00	Major,
	2	1		10 00	Adjutant, in addition to pay, etc., of Lieutenant,

TABLE OF PAY, SUBSISTENCE, FORAGE, ETC., OF THE U. S. ARMY.—*Continued.*

GRADE.	Pay. Per month.	Subsistence. No. of rations per day.	Forage. No. of horses allowed in time of war.	Forage. No. of horses allowed in time of peace.	No. of servants allowed.
Regimental Quartermaster, in addition to pay, etc., of Lieutenant,	$10 00				
Captain,	60 00	4	2	2	1
First Lieutenant,	50 00	4			1
Second Lieutenant,	45 00	4			1
Cadet,	24 00				
Sergeant-Major,	21 00				
Quartermaster-Sergeant,	21 00				
Principal Musician of Infantry,	21 00				
First Sergeant,	20 00				
Ordnance-Sergeant, in addition to pay of Sergeant,	5 00				
Sergeant,	17 00				
Corporal,	13 00				
Artificer of Artillery,	15 00				
Musician,	12 00				
Private,	11 00				

SAPPERS, MINERS AND PONTONIERS.

Sergeant,	.	34 00
Corporal,	.	20 00
Musician,	.	12 00
Private of the 1st class,	.	17 00
Private of the 2d class,	.	13 00

The commanding officer of a company is entitled to $10 per month for responsibility of arms and clothing.

Officers' subsistence is commuted at thirty cents per ration, forage, $8 per month for each horse actually owned and kept in service.

Officers are entitled to the pay of private soldier, $2 50 per month, clothing, and one ration per day for each private servant actually employed.

Every commissioned officer below the rank of Brigadier-General is entitled to one additional ration per day for every five years' service.

One dollar per month is to be retained from the pay of each private soldier until the expiration of his term of enlistment.

All enlisted men are entitled to $2 per month additional pay for re-enlisting, and $1 per month for each subsequent period of five years' service, provided they re-enlist within one month.

Paymasters' clerks, $7 00 per annum, and 75 cents per day when actually on duty.

* Entitled to only three rations per day as Lieutenant.
† Only the Captains and subalterns of dragoons, cavalry and mounted riflemen are entitled to *two* horses in time of peace.

ORDNANCE.

THE frequent mention of the terms "Mortars," "Dahlgren gun," "Columbiads," "Armstrong gun," "Howitzers," etc., renders a notice of these several "man killers" proper.

A mortar is a short cannon, bell-shaped, and is used principally for throwing shells filled with explosive materials for crushing and destroying buildings in sieges. They were used in Europe four centuries ago, but their destructive powers have acquired such terrific energy from modern improvements, that the bombardment of a city, or shelling it, is perhaps the most horrible fate to which it can be subjected. The bursting of a single shell spreads havoc and death among all near whom it may explode. It was these terrible missives which wrought the complete destruction of the quarters in Fort Sumter. Colonel Anderson's extreme care of his men on that occasion alone saved them from death, by keeping look-outs to give warning of the approach of shot or shell, and requiring his brave men to take shelter within the bomb-proof portions of the fort. Shells were rained down upon Sevastopol in such awful showers that the Russian commander wrote to the Emperor that it was the "fire of hell itself." The shell is supplied with a fuse, which takes fire when the mortar is exploded, and reaches the powder within the shell very often at the moment of its striking. Long practice has enabled gunners to know exactly the length of fuse necessary to the distance which the shell is to traverse. Many mortars in our service will throw a shell thirteen inches in diameter. Shells were formerly called bombs, and hence the word bombardment.

A howitzer is a gun with a chamber in it, and is used generally to throw shells and other hollow projectiles, which act as well by their explosion as by their force of percussion, setting fire to towns, ships and other quarters of an enemy. The field-howitzer is of course a much lighter gun than the siege or garrison-howitzer, and is used in light batteries. The mountain-howitzer is a very light twelve-pounder, and is used for service in countries so rough as not to admit the passage of wheeled vehicles. The howitzer and its carriage, when taken to pieces, are carried on the backs of mules, which, when the roads are favorable, may be used to draw the common two-wheeled carriage, with the mounted piece. The howitzer is lighter and shorter, in proportion to its projectile, than the ordinary cannon; the charges used are smaller, and the accuracy of fire much less. But this is compensated for by the greater execution of the shell when it bursts. The system of shell-guns was first brought into practical use by the French General, Paixan, in 1822, soon after which it was adopted by the United States. What is known among us as a columbiad is, in reality, a modification of the Paixan gun. There are two sizes, carrying balls eight and ten inches in diameter, either hollow or solid. This gun, therefore, combines the essential qualities of the ordinary cannon, the howitzer and the mortar. It discharges shot or shell with much greater precision than a mortar, and is terribly destructive. In casting columbiads, it has been found almost impossible to make them strong enough to withstand the proper number of discharges. It is a remarkable fact that the length of time that a piece has been cast has much influence on its power of endurance. On trial of three eight-inch columbiads, cast in the same mold and at the same time, one of them, a few days after casting, burst at the seventy-second round. Of the other two, after lying six years, one burst after eight hundred rounds, the other sustained two thousand five hundred and eighty-two fires without yielding. It is considered that all iron guns, after one thousand two hundred rounds, are no longer safe.

The Armstrong gun is a rifled cannon of English invention, and is loaded at the breech. All accounts agree in presenting it as throwing balls further, and more accurately than any other gun. It throws an explosive ball filled with percussion powder, which explodes when the ball strikes, tearing to pieces every thing near it. The British have introduced it extensively into the army. In their recent campaign in China it proved to be a terrific engine of death. At six hundred yards a target no larger than a man's hat, has been struck at almost every discharge, and at three thousand yards an object nine feet square, which can barely be seen at that distance, can be struck by every other shot. The largest Armstrong guns yet made carry balls weighing one hundred and twenty pounds. The number of guns made last year was seven hundred and eighty, at an average cost of one thousand seven hundred and fifty dollars. There are very few such guns in this country, perhaps none.

The steam-frigate Niagara is armed with sixty Dahlgren guns, weighing nearly five tons each. Six of them carry a ten-inch ball or shell, the remainder nine-inch, and will throw them from two to three miles. This gun was contrived by Captain Dahlgren of our navy, after a multitude of trials to determine the best form to avoid bursting. As all guns burst at or very near the breech, this is made of extraordinary thickness at that part, and for some three feet of its length, when it tapers down sharply to the muzzle. One of these guns has been made which weighs sixteen thousand pounds, and will throw an eleven-inch shell four miles.

The highest engineering talent, both in Europe and in this country, has been devoted to the construction of improved ordnance. Science and mechanical ingenuity, combined with millions of money, have been devoted to the business of discovering and remedying the defects of these death-dealing engines, as well as of inventing others more destructive. The French have rendered the rifle ten times as deadly as formerly, while the English, in the Armstrong gun, seemed to have rendered the cannon as true a marksman as the rifle. Up to this time, experience has demonstrated that there are insuperable obstacles to the formation of cast-iron guns of more than ten-inch caliber. Beyond that, so many defects are liable to occur in the casting, it is not considered safe to go, though with mortars the caliber has been extended to thirteen inches.

As all attempts to make casting larger than this have failed, attention has been turned to wrought-iron, with the hope of increasing the caliber. Yet the use of wrought-iron has been proved to be almost as dangerous as that of cast-iron. It is liable to many destructive casualties even after perfect welding has been secured. The huge wrought-iron gun which burst on board the Princeton in 1843, was proved to have parted with one-third of the original strength of the iron by the intense heat used in forging it. Wrought-iron guns of large caliber must be built up of separate pieces, and only a small amount of heat and welding force employed.

All the ordnance for the United States service is made at private founderies, and afterward inspected and proved by officers of the ordnance detailed for that purpose. The founderies where most of our cannon are made, are the following: The West Point Foundery, near Cold Spring, N. Y.; Fort Pitt, near Pittsburg; Tredegar, near Richmond, Va.; Algers, near Boston; and the Ames Foundery, near Chicopee, Mass. The last two furnish the bronze cannon, and the others the iron. By the term *artillery*, is meant all fire-arms of large caliber, together with the machines and implements used with them. In the United States service the artillery corps is intrusted with the use of the arms and munitions, and the *ordnance* corps with their construction and preservation. The term *ordnance* is applied to the guns themselves, and in our service the ordnance is divided into guns, howitzers and mortars.

RIFLES AND THEIR USE.

THIS weapon is now becoming a great favorite with all classes of regiments. Even the "Zouaves" are calling for the Minié or Sharpe's rifle. The musket seems to be regarded as less effective and sure in its mission of death. Notwithstanding, it is "Uncle Sam's favorite," and we think will tell as true a story when a campaign is over as the more expensive rifle—an instrument that needs great practice to render it available and effective. Hence, schools of practice, established by Government, are indispensable, and should be organized in this country. The French have such at Vincennes, Toulouse, St. Omer and Grenoble, whence officers and men, well instructed in the principles of firing, are sent out into the army at large, and impart to it the same system and efficiency. An attempt was made to establish such a school at Fórt Monroe for artillery practice, but schools for infantry are far more pressingly needed. The value of the bayonet in battle is well understood; but the fact is now incontestible that the efficiency of a body of infantry resides essentially in its accuracy of fire, a fact made more apparent from the recent improvements in fire-arms. A cool, well-directed fire from a body of men, armed with the best modern rifle or rifle-musket, is sufficient to stop the advance of almost any body of troops, but the very best disciplined men will, in time of battle, fire with precipitancy and at too great a distance. The thoroughness of practice, also, should be in proportion to the efficiency of the troops to be encountered. In one of Colonel Steptoe's encounters with the Indians of Washington Territory, his men were armed with the old musket, and they soon expended their ammunition in ineffectual firing against enemies mounted on fleet horses, armed partly with rifles, partly with bows and arrows, whose deadly shaft was shot with astonishing accuracy, and at a rate exceeding the rapidity of an expert hand with a revolver. Charges of cavalry against them failed, and our men retreated to avoid annihilation. Some weeks subsequently the same troops met the same Indians, but having in the interval procured the rifle instead of the musket, the Indians were totally routed.

In the successful and oft-repeated repulse of cavalry charges by squares of infantry, the main dependence is not on the use of the bayonet; but in the close, well-directed fire, delivered as the horsemen approach. This, breaking their formation, and disorganizing their ranks, leaves them at the end of their charge with a wall of bayonets in front, against which horses can not be forced unless at full speed and supported by numbers behind. It is this injury, before the shock takes place, which prevents cavalry from breaking squares of infantry. So accurate and fatal has the rifle become by modern improvement, that it has been customary to underrate the artillery arm on the field of battle; and the assertion is frequently made that the use of the rifle will entirely supersede the use of field-pieces in war, since their fire has a greater range and more accuracy than the field-pieces now in use. But able military writers doubt this. Others insist that artillerymen will be shot down at such a distance from their guns as to make it impossible to serve them in the face of infantry; that bayonets will not be crossed so often; that personal conflicts, such as line against line, or column against column, will cease altogether, and future combats be decided by the effects of a rapid and destructive fire, on the precision of which, rather than on personal contact and extensive combinations, the result will depend. In India, Havelock mowed down whole columns of advancing insurgents by Minié rifles. They worked dreadful havoc to the Austrians at Magenta and Solferino. Garibaldi retreated before their withering fire at Rome. At the battle of Ilstedt, a body of skirmishers, armed with rifles, discharging conical balls, made an attack on the Danes at a distance of one hundred and fifty yards. Artillery replied to them, cavalry made repeated charges at them, and infantry advanced, but they could not be moved. In less than an hour they killed seventy men, with several officers of high rank, and ninety horses. This havoc gives color to the idea of the improved rifle superseding field-artillery.

The fire of the ordinary musket is uncertain beyond two hundred yards, but when troops are in compact masses, it is still very effective beyond that distance. At six hundred and fifty yards the musket-ball is still deadly, and has been known to kill at even greater distances. The effective range of the rifled spherical ball is over four hundred yards; the oblong rifle-ball is effective at one thousand yards, or more than half a mile. The rifle was in use as early as 1498. Almost every European army has adopted its own kind of rifle, essentially differing from each other. In the small German States they use bullets of almost every conceivable shape and size. The principal arms adopted for the British army is the Enfield rifled musket, manufactured in the Government establishment. Its principal competitor is the Whitworth rifle. The French are now perhaps more advanced than all others in experience and efficiency with that arm. To Captain Minié, of their army, they are indebted for the weapon bearing his name, and it is now being rapidly adopted throughout the French service.

The schools for practice established in that country should be imitated here. The men are there taught to take the easiest and most stable positions, either standing or kneeling; to sight and fire with blank cartridges, preserving immovable both the body and the piece. The quick movement the soldier imparts to the piece, by pulling the trigger, is the great cause of his losing his aim. The principal object of instruction is to habituate him not to being surprised by the explosion, by pressing gradually upon it. For this purpose they are made to fire caps at a lighted candle, placed about three inches from the piece. If the piece is properly aimed, the jet of gas produced by the cap will extinguish the candle. After this, they fire blank cartridges. The officers are thoroughly instructed in estimating distances, as they will so much better direct the fire of the men they command. This knowledge is of great advantage to them in maneuvering troops, and soldiers thrown out as skirmishers will out-general an enemy if they know how to estimate distances with precision, for their fire will then be more accurate and efficacious. There are regular instructors in all these departments. Records of the firing are kept in each battalion. It is to this thorough training, combined with the use of the best weapon, that the French army owes most of its present supremacy. The men are as much marksmen as any of our deer-slayers. The Minié ball, fired by them, is a terrific missile. What destruction it works was shown at the recent riot at St. Louis. Those which struck the walls of the houses, tore up bricks for a space of four inches in diameter. When they struck fair, they sunk six inches into the solid wall. One of them struck the angle of a wall, tore away a brick next to the door-frame, passed six inches through the frame, then through the door and into the wall beyond. A stroke of lightning has frequently done less damage.

At six hundred yards, and even further, a handful of riflemen can render field-artillery useless in a few minutes, by the destruction of its ammunition and gunners; and even a single rifleman, well ensconced, may silence a field-battery : a feat performed by Lieutenant Godfrey of the Rifles, who silenced a Russian battery at Balaklava. A bullet has been recently invented by the commander of the Normal School of Gunnery, in France, which, besides superiority of precision over that of the rifle-ball, may be fired from a smooth barrel. Of one hundred balls fired from a rifle, at six hundred yards distance, forty-three struck the target, and at eight hundred yards, fourteen. Of one hundred of the newly-invented ball fired from a smooth barrel, at the distance of six hundred yards, sixty-five struck the target, and at eight hundred yards, thirty-nine.

CULINARY DEPARTMENT.

Preparation of Food.

To prepare good food from the ordinary ration is comparatively an easy matter; but, so poorly do most persons understand the art of compounding, or even of simple roasting, boiling and dressing, that soldier's fare has become proverbial for its unpalatableness. It was no matter of surprise to the French, in the late War in the Crimea, to see their allies, the English, die off by hundreds, when their detestable arrangements in the commissary department and hospitals were unveiled; and it was not until the English government sent skillful cooks (headed by the celebrated Soyer), to show the troops *how to live*, that the awful mortality ceased. Soyer found the army eating half-cooked beans, raw salt meat and hard bread rather than be at the trouble of preparing their rations by careful cooking. He instituted a thorough *system* in the culinary department, and, by his excellent arrangements and thoroughly scientific principles of preparing and mixing the food, produced a most astonishing change in the comfort, health and happiness of the entire army. His published " Cookery for the Army" places us in possession of the hints and recipes requisite for a *good table* from the ordinary army ration list, and we here subjoin such of the recipes as our troops may render available for their comfort and health :

To Cook Salt Meat for Fifty Men.*—Put fifty pounds of meat in the boiler. Fill with water, and *let soak all night*. Next morning wash the meat well. Fill with fresh water, and boil gently three hours, and serve. Skim off the fat, which, when cold, is an excellent substitute for butter.

For salt pork proceed as above, or boil half beef and half pork—the pieces of beef may be smaller than the pork, requiring a little longer time doing.

Dumplings may be added to either pork or beef in proportion ; and when pork is properly soaked, the liquor will make a very good soup, by the addition of five pounds of split peas, half a pound of brown sugar, two table-spoonfuls of pepper, ten onions ; simmer gently till in

* In all these cases, dishes for half or quarter of the number named, use only half or quarter of the proportions given.

pulp, remove the fat and serve; broken biscuit may be introduced. This will make an excellent mess.

SALT PORK WITH MASHED PEAS, FOR ONE HUNDRED MEN.—Put in two stoves fifty pounds of pork each; divide twenty-two pounds, in four pudding-cloths, rather loosely tied, putting to boil at the same time as your pork; let all boil gently till done, say about two hours; take out the pudding and peas; put all meat in one caldron; remove the liquor from the other pan, turning back the peas in it; add two teaspoonfuls of pepper, a pound of the fat, and with the wooden spatula smash the peas and serve both. The addition of about half a pound of flour and two quarts of liquor, boiled ten minutes, makes a great improvement. Six sliced onions, fried and added to it, makes it very delicate.

STEWED SALT BEEF AND PORK FOR ONE HUNDRED MEN.—Put in a boiler, of well-soaked beef thirty pounds, cut in pieces of a quarter of a pound each; twenty pounds of pork; one and a half pound of sugar; eight pounds of onions, sliced; twenty-five quarts of water; four pounds of rice. Simmer gently for three hours, skim the fat off the top, and serve.

SOUP FOR FIFTY MEN.—Put in the boiler sixty pints, seven and a half gallons, or five and a half camp-kettles of water. Add to it fifty pounds of meat, either beef or mutton; the rations of preserved or fresh vegetables; ten small table-spoonfuls of salt. Simmer three hours, and serve. When rice is issued put it in when boiling. Three pounds will be sufficient. About eight pounds of fresh vegetables; or four squares from a cake of preserved ditto; a table-spoonful of pepper, if handy. Skim off the fat, which, when cold, is an excellent substitute for butter.

SOUPS FOR TWO MEN*—*Camp Soup.*—Put half a pound of salt pork in a saucepan, two ounces of rice, two pints and a half of cold water, and, when boiling, let simmer another hour, stiring once or twice; break in six ounces of biscuit; let soak ten minutes; it is then ready, adding one tea-spoonful of sugar, and a quarter one of pepper, if handy.

Beef Soup.—Proceed as above, boil an hour longer, adding a pint more water.

[Those who can obtain any of the following vegetables will find them a great improvement to the above soups : Add four ounces of either onions, carrots, celery, turnips, leeks, greens, cabbage or potatoes, previously well washed or peeled, or any of these mixed to make up four ounces, putting them in the pot with the meat. The green tops of leeks and the leaf of celery as well as the stem, for stewing are preferable to the white part for flavor. The meat being generally salted with rock

---

* For more than two men simply increase the proportions *pro rata.*

salt, it ought to be well scraped and washed, or even soaked in water a few hours if convenient; but if the last can not be done, and the meat is therefore too salt, which would spoil the broth, parboil it for twenty minutes in water before using for soup, taking care to throw this water away.]

Fresh Beef Soup.—For fresh beef proceed, as far as the cooking goes, as for salt beef, adding a tea-spoonful of salt to the water.

Bean or Pea Soup.—Put in your pot half a pound of salt pork, half a pint of peas, three pints of water, one tea-spoonful of sugar, half one of pepper, four ounces of vegetables cut in slices, if to be had; boil gently two hours, or until the peas are tender, as some require boiling longer than others, and serve.

Fresh Beef Soup, or Pot-au-feu—Camp Fashion, for the Ordinary Canteen Pan.—Put in the canteen saucepan six pounds of beef, cut in two or three pieces, bones included, three-quarter pound of plain mixed vegetables—as onions, carrots, turnips, celery, leeks, or such of these as can be obtained—or three ounces of preserved in cakes, as now given to the troops; three tea-spoonfuls of salt, one ditto of pepper, one ditto of sugar, if handy; eight pints of water; let it boil gently three hours; remove some of the fat, and serve. The addition of one and a half pound of bread cut into slices, or one pound of broken biscuits, well soaked, in the broth, will make a very nutritious soup; skimming is not required.

Plain Irish Stew for Fifty Men.—Cut fifty pounds of mutton into pieces of a quarter of a pound each; put them in the pan; add eight pounds of large onions, twelve pounds of whole potatoes, eight table-spoonfuls of salt, three table-spoonfuls of pepper; cover all with water, giving about half a pint to each pound; then light the fire; one hour and a half of gentle ebulition will make a most excellent stew; mash some of the potatoes to thicken the gravy, and serve. Fresh beef, veal, or pork, will also make a good stew. Beef takes two hours doing. Dumplings may be added half an hour before done.

Semi-Frying, Camp Fashion, Chops, Steaks, and All Kinds of Meat.—If it is difficult to broil to perfection, it is considerably more so to cook meat of any kind in a frying-pan. Place your pan on the fire for a minute or so; wipe it very clean; when the pan is very hot, add in it either fat or butter, but the fat from salt and ration meat is preferable; the fat will immediately get very hot; then add the meat you are going to cook; turn it several times to have it equally done; season to each pound a small tea-spoonful of salt, quarter that of pepper, and serve. Any sauce or maitre d'hotel butter may be added. A few fried onions in the remaining fat, with the addition of a little flour to the onions, a quarter of a pint of water, two table-spoonfuls of vinegar, a few chopped pickles or piccalilly, will be very relishing.

RECEIPTS FOR THE FRYING-PAN.—Those who are fortunate enough to possess a frying-pan will find the following receipts very useful: Cut in small dice half a pound of solid meat, keeping the bones for soup ; put your pan, which should be quite clean, on the fire ; when hot through, add an ounce of fat, melt it and put in the meat, season with half a tea-spoonful of salt; fry for ten minutes, stirring now and then ; add a tea-spoonful of flour, mix all well, put in half a pint of water, let simmer for fifteen minutes, pour over a biscuit previously soaked, and serve.

The addition of a little pepper and sugar is an improvement, as is also a pinch of cayenne, curry-powder, or spice ; sauces and pickles used in small quantities would be very relishing; these are articles which will keep for any length of time. As fresh meat is not easily obtained, any of the cold salt meat may be dressed as above, omitting the salt, and only requires warming; or, for a change, boil the meat plainly, or with greens, or cabbage or dumplings, as for beef; then the next day cut what is left in small dice—say four ounces—put in a pan an ounce of fat; when very hot, pour in the following: Mix in a basin a table-spoonful of flour, moisten with water to form the consistency of thick melted butter, then pour it in the pan, letting it remain for one or two minutes, or until set; put it in the meat, shake the pan to loosen it, turn it over, let it remain a few minutes longer, and serve.

To cook bacon, chops, steaks, slices of any kind of meat, salt or fresh sausages, black puddings, etc. Make the pan very hot, having wiped it clean, add in fat, dripping, butter, or oil, about an ounce of either ; put in the meat, turn three or four times, and season with salt and pepper. A few minutes will do it. If the meat is salt, it must be well soaked previously.

TURKISH PILAFF FOR ONE HUNDRED MEN.—Put in the caldron two pounds of fat, which you have saved from salt pork, add to it four pounds of peeled and sliced onions ; let them fry in the fat for about ten minutes; add in then twelve pounds of rice; cover the rice over with water, the rice being submerged two inches; add to it seven table-spoonfuls of salt and one of pepper ; let simmer gently for about an hour, stirring it with a spatula occasionally to prevent it burning, but when commencing to boil, a very little fire ought to be kept under. Each grain ought to be swollen to the full size of rice and separate. In the other stove put fat and onions the same quantity, with the same seasoning ; cut the flesh of the mutton, veal, pork, or beef from the bone, cut in dice of about two ounces each, put in the pan with the fat and onions, set it going with a very sharp fire, having put in two quarts of water ; steam gently, stirring occasionally for about half an hour, till forming rather a rich thick gravy. When both the rice and meat is done, take half the rice and mix with the meat, and then the remainder of the meat and rice, and serve; save the bones for soup the following day. Salt pork or beef well soaked, may be used—omitting the salt. Any kind of vegetables may be frizzled with the onions.

COFFEE A LA ZOUAVE FOR A MESS OF TEN SOLDIERS.—Make it in the canteen saucepan holding ten pints. Put nine pints of water into a canteen saucepan on the fire; when boiling, add seven and a half ounces of coffee, which forms the ration, mix them well together with a spoon or a piece of wood, leave on the fire for a few minutes longer, or until just beginning to boil. Take it off and pour in one pint of cold water; let the whole remain for ten minutes or a little longer. The dregs of coffee will fall to the bottom, and your coffee will be clear. Pour it from one vessel to the other, leaving the dregs at the bottom; add your ration sugar, or two tea-spoonfuls to the pint; if any milk is to be had make two pints of coffee less; add that quantity of milk to your coffee; the former may be boiled previously, and serve.

This is a very good way for making coffee, one ounce to the quart if required stronger. For a company of eighty men use four times the quantity of ingredients.

COFFEE, TURKISH FASHION.—When the water is just on the boil add the coffee and sugar, mix well as above, give just a boil, and serve. The grouts of coffee will in a few seconds fall to the bottom of the cups. The Turks wisely leave it there. I would advise every one in camp to do the same.

COCOA FOR EIGHTY MEN.—Break eighty portions of ration cocoa in rather small pieces; put them in the boiler, with five or six pints of water; light the fire, stir the cocoa round till melted, and forming a pulp not too thick, preventing any lumps forming; add to it the remaining water, hot or cold; add the ration sugar, and when just boiling, it is ready for serving. If short of cocoa in campaigning, put about sixty rations, and when in pulp, add half a pound of flour or arrow-root.

TEA FOR EIGHTY MEN.—One boiler will, with ease, make tea for eighty men, allowing a pint each man. Put forty quarts of water to boil, place the rations of tea in a fine net very loose, or in a large perforated ball; give one minute to boil, take out the fire, if too much, and shut down the cover; in ten minutes it is ready to serve.

This will do for Soyer, which dishes, after all, appear to us to be *too expensive.* For instance, in the preparation of his soups he gives *one pound of meat* to every man to be fed. As the men are only rationed with three-quarters of a pound of pork *per day,* or twenty ounces of salt or fresh beef, the entire daily ration is thus consumed to prepare one dish. We never, in our own household, use so great a proportion of meat in soups, while other dishes are prepared by good housewives with more economy than those given above. We shall, therefore, here add a number of recipes *especially* prepared for this work by an able and experienced American housewife, which, we hope, will prove very acceptable and valuable to the company cook:

Hoe-Cake.—Mix a stiff dough of Indian meal, a little salt, and water (scalding water is best); flatten it on a board, and tilt it up before the camp-fire until brown on one side; turn, and brown the other. When our fathers fought the Indians, and ground their corn in mortars, they thought hoe-cake very good. It can also be baked in hot ashes, and with hot stones, *Southern* fashion.

Short-Cake.—Mix a dough of flour, salt, and cold water, and bake in the same way. If you have any kind of fat, rub a little into the flour thoroughly, before adding the water. If you have baking-powder, use it, a tea-spoonful to a pint of flour, stirring it in before adding the water; but it is good without, when baked before the fire.

Broiled Meat.—When you have fresh meat, broil it, in preference to frying it. It is much sweeter, and far healthier. A bit of fresh meat, stuck on a stick (in the absence of a gridiron), and broiled over the coals, is more delicious than the most carefully prepared dish.

Salt Pork—Is also excellent, broiled before the fire or on the coals. Put slices on the end of a stick sharpened at both ends and set aslant before the fire, so that the fat will drop off, on a piece of bread or biscuit which can be placed beneath it, and which will brown at the same time, and make an acceptable relish.

Hash.—Cut up the cold bits of corned beef with the cold potatoes (if you have no potatoes, dry bread or biscuit will answer very well), about half and half; heat a kettle or tin-dish with a bit of fat in the bottom, put in the hash, moisten with water, season with pepper and salt, stir till well heated through. A good dish for breakfast, and makes the "pieces" available.

Baked Beans.—If on the march, and you have not time to bake beans for dinner, you can have them for breakfast. Let them boil until soft, in two waters. In the first water put a tea-spoonful of soda to a quart of beans; when they have cooked in this twenty or thirty minutes, pour it off, and add another water, enough for them to swell in until soft. Do this in the evening; before bed-time, put the beans in a pan, with a chunk of pork in the middle, and a little salt and pepper to season, and let them stand where they will bake *slowly* through the night—in a slow oven, if you have a cooking-stove; if not, in hot ashes, with some coals or heated stones on the cover of the pan. They will be ready for breakfast; and this is the real Boston way of eating beans of a Sunday morning.

BOILED BEANS.—When there is no chance to soak beans over night, put soda in the first water, turn it off, after boiling half an hour, add fresh water, and boil two hours, with a chunk of pork.

COFFEE.—Don't spoil your coffee by pouring the water on it *before it boils*, or by adding water to it, *after* it is steeped. Mix the ground coffee first with cold water to a thin paste; pour on boiling water, allowing a table-spoonful of coffee to a person, and a pint of water ; boil quickly ten minutes, let it stand to settle three or four minutes, and pour off from the dregs. If you have egg-shells, or a bit of codfish-skin, mix with the ground coffee before adding water; they will help to make it clear. If you have milk, *boil it*, and add it to the coffee after it is poured from the dregs.

TEA.—A tea-spoonful to a person. Always use *boiling* water; steep two or three minutes ; then let it stand to "draw" as many more. If you have any tea left, do not throw it away. Fill your canteens with it. It is infinitely more refreshing than almost any other drink, upon a hot, weary march. If, instead of filling your canteens with fresh water, you would *boil* it in the morning, before starting, with enough tea to flavor it and keep it from being insipid when warmed by the sun, it would be a thousand times healthier, and the best preventive against dysentery. Water which has been boiled is freed from the bad effects which it frequently has. The Southern people *boil their lemonade*, and then allow it to cool, before using it. Learn from your enemies how to protect yourselves in their climate.

EGGS—May be roasted by standing them on end in hot ashes ; they may be boiled hard to carry in the pockets on forced marches ; they may be scrambled, by melting a piece of fat or butter the size of an egg to a dozen eggs in a hot dish of any kind you have, frying-pan, kettle, tin-basin, etc., breaking the eggs in and stirring them about till they are thick, adding pepper and salt, also chopped ham, if you have it. To make a dozen eggs go as far as two dozen, heat the pan as above, with a piece of fat, cut bread in square dice, put in the pan and fry a light brown, then add the eggs, and stir about till set. Or beat a couple of eggs in half a pint of milk, dip slices of bread in the mixture and fry a light brown.

SOUP.—Unless compelled, by being on the march, never throw away the bones of any meat, fresh or salt, nor the pieces—they make good soup, with the addition of very little or no meat, and thus save the rations of meat, so that the soldier can sometimes have *two courses* for his dinner. Of course when there is less meat, there must be more thickening with other substances. Almost any kind of vegetables, chopped or sliced, barley, rice, or even flour, *browned in a hot frying-pan*, and cooked well into the soup, and well seasoned with pepper and salt, will make palatable soup, with but a little meat to flavor it. Onions are especially useful in flavoring soup. Peas and bean soup needs less meat, because the vegetables themselves are nourishing. They should be boiled until thoroughly dissolved, with a chunk of pork, or a ham-bone. An excellent thickening for soups may be made, by mixing flour and fat, and frying brown, and stirring into the soups a half-hour before they are taken off. Vegetables should be put into soup from two hours to an hour before it is done; rice or flour an hour before, pepper or other pungent flavorings just before it is taken from the fire.

DUMPLINGS—May be made to add to soup, when there is a lack of vegetables, and to all stewed meats, for the same reason. Mix a dough, as for biscuits, by rubbing a small piece of fat into a quart of flour; stir two tea-spoonsful of baking-powder through it, mix it up, not very hard, with cold water, cut into slices, or roll into small balls, and boil three-quarters of an hour in soup, or along with meat which is stewing.

WARMED-UP MEAT FOR BREAKFAST.—Cut cold corned-beef in slices or small pieces. Put a piece of fat or butter, the size of an egg, in a frying-pan, pour on a pint of water, mix a table-spoonful of flour to a paste with cold water, thicken the gravy with this, and warm up the beef in the gravy.

RICE.—To boil rice, put it in cold water, let it swell upon a slow fire, until tender, without stirring; then add salt. It can be eaten with sweet gravy, with butter and pepper, with sugar or molasses, with sugar and butter, or plain. Take the cold rice and fry it for supper or breakfast, by cutting it in slices, or making it in balls, and frying it on both sides in a little fat.

PEPPER.—If you will learn to use cayenne pepper in place of black you will find it a preventive of dysentery, and a cure for colds. An extra pinch of it in your breakfast will often break up a cold caught through the night; and a smart sprinkle of it in your liquid will relieve sickness caused by bad drinking-water. Black pepper produces inflammation—red pepper heals it.

" If the voice of a woman could prevail, she would have red pepper take the place of black in the army ; and have all drinking-water *boiled* before using, whenever possible ; and further, have it flavored with coffee, or tea—tea being preferable. It would promote the cause of temperance and be a most effective sanitary measure." So writes our good house-wife. Is there not good sense in her suggestions?

Other Good Things.

From Soyer's recipes we may quote the following as embodying many good things which the soldier can have prepared for his "occasional treat:"

CHEAP PLAIN RICE PUDDING, FOR CAMPAIGNING—In which no eggs or milk are required; important in the Crimea or the field. Put on the fire, in a moderate sized saucepan, twelve pints of water; when boiling, add to it one pound of rice or sixteen table-spoonfuls, four ounces of brown sugar or four table-spoonfuls, one large tea-spoonful of salt, and the rind of a lemon thinly peeled; boil gently for half an hour, then strain all the water from the rice, keeping it as dry as possible. The rice water is then ready for drinking, either warm or cold. The juice of a lemon may be introduced, which will make it more palatable and refreshing.

THE PUDDING.—Add to the rice three ounces of sugar, four table-spoonfuls of flour, half a tea-spoonful of pounded cinnamon; stir it on the fire carefully for five or ten minutes; put it in a tin or a pie-dish and bake. By boiling the rice a quarter of an hour longer, it will be very good to eat without baking. Cinnamon may be omitted.

BATTER PUDDING.—Break two fresh eggs in a basin, beat them well, add one table-spoonful and a half of flour, which beat up with your eggs with a fork until no lumps remain; add a gill of milk and a tea-spoonful of salt; butter a tea-cup or a basin, pour in your mixture, put some water in a stewpan, enough to immerge half way up the cup or basin in water; when boiling, put in your cup or basin, and boil twenty minutes, or till your pudding is well set; pass a knife to loosen it, turn out on a plate, pour pounded sugar and a pat of fresh butter over, and serve. A little lemon, cinnamon, or a drop of any essence may be introduced. A little light-melted butter, sherry and sugar, may be poured over. If required more delicate, add a little less flour. It may be served plain.

BREAD AND BUTTER PUDDING.—Butter a tart-dish well, and sprinkle some currants all around it, then lay in a few slices of bread and butter; boil one pint of milk, pour it on two eggs well whipped, and then on the bread and butter; bake it in a hot oven for half an hour. Currants may be omitted.

BREAD PUDDING.—Boil one pint of milk, with a piece of cinnamon and lemon-peel; pour it on two ounces of bread-crumbs; then add two eggs, half an ounce of currants, and a little sugar; steam it in a buttered mold for one hour.

CUSTARD PUDDING.—Boil one pint of milk, with a small piece of lemon-peel and half a bay leaf, for three minutes; then pour these on to three eggs, mix it with one ounce of sugar well together, and pour it into a buttered mold; steam it twenty-five minutes in a stewpan with some water; turn out on a plate and serve.

RICH RICE PUDDING.—Put half a pound of rice in a stewpan, washed, three pints of milk, one pint of water, three ounces of sugar, one lemon-peel, one ounce of fresh butter; boil gently half an hour, or until the rice is tender; add four eggs, well beaten; mix well, and bake quickly for half an hour, and serve. It may be steamed if preferred.

STEWED MACARONI.—Put in a stewpan two quarts of water, half a table-spoonful of salt, two ounces of butter; set on the fire; when boiling, add one pound of macaroni, broken up rather small; when boiled very soft, throw off the water; mix well into the macaroni a table-spoonful of flour, and add enough milk to make it of the consistency of thin melted butter; boil gently twenty minutes; add in a table-spoonful of either brown or white sugar, or honey, and serve. A little cinnamon, nutmeg, lemon-peel, or orange-flower water may be introduced to impart flavor; stir quick. A gill of milk or cream may now be thrown in three minutes before serving. Nothing can be more nutritious than macaroni done this way. If no milk, use water.

MACARONI PUDDING.—Put two pints of water to boil; add to it two ounces of macaroni, broken in small pieces; boil till tender, drain off the water, and add half a table-spoonful of flour, two ounces of white sugar, a quarter of a pint of milk, and boil together for ten minutes; beat an egg up, pour it to the other ingredients, with a nut of butter; mix well and bake, or steam. It can be served plain, and may be flavored with either cinnamon, lemon, or other essences, as orange-flower water, vanilla, etc.

SAGO PUDDING.—Put in a pan four ounces of sago, two ounces of sugar, half a lemon-peel or a little cinnamon, a small pat of fresh butter, if handy, and half a pint of milk; boil for a few minutes, or until rather thick, stirring all the while; beat up two eggs and mix quickly with the same; it is then ready for either baking or steaming, or may be served plain.

TAPIOCA PUDDING.—Put in a pan two ounces of tapioca, one and a half pint of milk, one ounce of white or brown sugar, a little salt; set on the fire; boil gently for fifteen minutes, or until the tapioca is tender, stirring now and then to prevent its sticking to the bottom, or burning; then add two eggs well beaten; steam or bake, and serve. It will take about twenty minutes steaming, or a quarter of an hour baking slightly. Flavor with either lemon, cinnamon, or any other essence.

HEALTH DEPARTMENT.

General Advice.

"A YOUNG trooper needs an old horse," says the proverb. The greater part of our volunteers are entering upon an untried life—a life in which they must rely upon themselves, individually, and in which their only guides are common sense and the experience of others. The first is not as common as it ought to be, and young men are proverbial for refusing to learn by the other. Yet there are many who are prudent enough to borrow a light, and to them a few hints from one accustomed to camp-life may be valuable. Without apology, they will be very plain and practical.

Every volunteer who goes from city or country to field and camp, and from home fare and regular habits to camp rations and military discipline, must go through a process of adaptation, more or less severe, even under the most favorable circumstances. In the present case a process of acclimation must be added, which will make the change unusually trying, and will require unusual prudence. A new army is never thrown into camp without incurring disease not necessarily severe, but which may be aggravated by carelessness into epidemic and chronic forms. Among others, diseases of the bowels are common, and almost inevitable. The first thing the young soldier will have to contend with is a craving appetite for something, which he is apt to gratify by a stimulant. If he is a temperate man he is likely to drink frequently and profusely of water. Both are to be resisted and avoided. The only possible good that spirits can do is in cases of great exhaustion, and even then their use is of *very exceptional utility*. A cup of coffee or tea is almost invariably better. The habit of drinking spirits in advance of labor, to gain strength, is one of the worst in the world. The action and reaction of a stimulant are equal, and the greater its strength the shorter its good effect, and the worse the depression afterward. If you *will* drink spirits, never do it *before* a march, or action, because the unnatural energy which it imparts may fail you at the very moment you need it most. As to water, nothing effects the bowels more certainly than liberal draughts of a kind to which we are unused. But we must drink something, you say. Not a quarter as much as you think. Drinking, in warm weather especially, is very much a matter of habit. Every one who exercises self-denial for a day or two will find that he really needs but little, and that, when he must drink, a single draught will do as well as a dozen.

A moderate use of salt rations, and especially of fat meat, will pay on the score of health, beyond calculation. Although there are occasions when a soldier's meals are necessarily irregular, it is a very favorable circumstance that, in camp, rations are usually served with perfect uniformity as to time. Every soldier should take all the time he can to eat, and be careful about indulging his appetite between meals, when he has access, as he sometimes will, to unused delicacies. In case of bowel complaints, if circumstances allow, fasting and lying flat upon the back (not upon the bare ground), and keeping as warm as possible, are usually more curative then medicines.

Nothing is more tempting, when a man is foot-weary after a hard march, then to plunge the feet into cold water. Nothing is more dangerous, or more likely to induce pulmonary diseases. Wait until the feet are *entirely dry and cool*—it is better to wait until the next morning. The same is true of all bathing—the face, wrists and hands only excepted—it should never be indulged when heated. Bathing early in the morning is healthful and invigorating, and should be regularly practiced, when possible, for the sake of cleanliness. A dirty man is always liable to disease. Yet it is not well to use much soap, except upon the hands, as the alkali contained in it unduly purges the pores. Water, the natural solvent, when freely used, is the most efficacious cleanser.

Next to the temptation of bathing at improper times, is that of throwing one's self on the cool, inviting ground, when hot and weary. Never do it, or at most for more than a minute. Leaning against a tree or the back of a comrade, as both sit upon blankets or knapsacks, is far better. When you must sleep on the ground, spread your india rubber blanket under, if you have one. Rheumatisms, agues, diarrheas, dysenteries, and fevers may thus be avoided.

Don't discard your flannel shirt because it is warm; and

always remember that a suddenly checked perspiration, whether by incautious draughts of water, by lying on the damp ground, by sitting in the wind, or otherwise, may be the means of swift disease and death. When you are on guard, or marching in the rain, keep your shoulders dry if you can. If you are wet through, keep going till your clothes dry, and you will not be likely to take cold.

Upon the recommendation of the Medical Commission, Governor Andrew has issued the following directions to the Massachusetts regiments in active service, to which all volunteers would do well to pay attention :

Special Directions.

Soldiers should recollect that in a campaign, where one dies in battle, from three to five die of disease. You should be on your guard, therefore, more against this than the enemy, and you can do much for yourselves which nobody can do for you.

Avoid especially all use of ardent spirits. If you will take them, take them rather *after* fatigue than before. But tea and coffee are much better. Those who use ardent spirits are always the first to be sick and the most likely to die.

Avoid drinking freely of very cold water, especially when hot or fatigued, or directly after meals. Water quenches thirst better when not very cold and sipped in moderate quantities slowly, though less agreeable. At meals, tea, coffee and chocolate are best. Between meals, the less the better. The safest in hot weather is molasses and water with ginger or small beer.

Avoid all excesses and irregularities in eating and drinking. Eat sparingly of salt and smoked meats, and make it up by more vegetables, as squash, potatoes, peas, rice, hominy, Indian meal, etc., when you can get them. Eat little between, when you have plenty at meals.

Wear flannel all over in all weathers. Have it washed often when you can ; when not, have it hung up in the sun. Take every opportunity to do the same by all your clothing, and keep every thing about your person dry, especially when it is cold.

Do not sit, and especially do not sleep upon the ground, even in hot weather. Spread your blanket upon hay, straw, shavings, brushwood, or any thing of the kind. If you sleep in the day, have some *extra covering* over you.

Sleep as much as you can, and whenever you can. It is better to sleep too warm than too cold.

Recollect that cold and dampness are great breeders of disease. Have a fire to get around whenever you can, especially in the evening and after rain, and take care to dry every thing in and about your person and tents.

Take every opportunity of washing the whole body with soap and water. Rinse well afterward. If you bathe, remain in the water but a little while.

If disease begins to prevail, wear a white bandage of flannel around the bowels.

Keep in the open air, but not directly exposed to a hot sun. When obliged to do this, a thin, light, white covering over the head and neck in the form of a cap with a cape, is a good protection.

Wear shoes with very thick soles, and keep them dry. When on the march, rubbing the feet after washing with oil, fat, or tallow, protects against foot-sores.

Doctors will differ somewhat in their views of the same subject; hence, we are not surprised to find that excellent authority, "Hall's Journal of Health," offering advice not entirely in consonance with the instructions of army surgeons. Still, the advice is so unquestionably *sensible* that we here give place to

Dr. Hall's Paper.

In any ordinary campaign, sickness disables or destroys three times as many as the sword.

On a march, from April to November, the entire clothing should be a colered flannel shirt, with a loosely-buttoned collar, cotton drawers, woolen pantaloons, shoes and stockings, and a light-colored felt hat, with broad brim to protect the eyes and face from the glare of the sun and from the rain, and a substantial but not heavy coat when off duty.

Sun-stroke is most effectually prevented by wearing a silk handkerchief in the crown of the hat.

Colored blankets are best, and if lined with brown drilling the warmth and durability are doubled, while the protection against dampness from lying on the ground, is almost complete.

Never lie or sit down on the grass or bare earth for a moment; rather use your hat—a handkerchief even, is a great protection. The warmer you are, the greater need for this precaution, as a damp vapor is immediately generated, to be absorbed by the clothing, and to cool you off too rapidly.

While marching, or on other active duty, the more thirsty you are, the more essential is it to safety of life itself, to rinse out the mouth two or three times, and *then* take a swallow of water at a time, with short intervals. A brave French general, on a forced march, fell dead on the instant, by drinking largely of cold water, when snow was on the ground.

Abundant sleep is essential to bodily efficiency, and to that alertness of mind which is all important in an engagement; and few things more certainly and more effectually prevent sound sleep than eating heartily after sun-down, especially after a heavy march or desperate battle.

Nothing is more certain to secure endurance and capability of long-continued effort, than the avoidance of every thing as a drink except cold water, NOT excluding coffee at breakfast. Drink as little as possible, of even cold water.

After any sort of exhausting effort, a cup of coffee, hot or cold, is an admirable sustainer of the strength, until nature begins to recover herself.

Never eat heartily just before a great undertaking; because the nervous power is irresistibly drawn to the stomach to manage the food eaten, thus drawing off that supply which the brain and muscles so much need.

If persons will drink brandy, it is incomparably safer to do so *after* an effort than before; for it can give only a transient strength, lasting but a few minutes; but as it can never be known how long any given effort is to be kept in continuance, and if longer than the few minutes, the body becomes more feeble than it would have been without the stimulus, it is clear that its use *before* an effort is always hazardous, and is always unwise.

Never go to sleep, especially after a great effort, even in hot weather, without some covering over you.

Under all circumstances, rather than lie down on the bare ground, lie in the hollow of two logs placed together, or across several smaller pieces of wood, laid side by side; or sit on

your hat, leaning against a tree. A nap of ten or fifteen
minutes in that position will refresh you more than an hour
on the bare earth, with the additional advantage of perfect
safety.

A *cut* is less dangerous than a bullet-wound, and heals more
rapidly.

If from any wound the blood spurts out in jets, instead of
a steady stream, you will die in a few minutes unless it is
remedied; because an artery has been divided, and that takes
the blood directly from the fountain of life. To stop this instantly, tie a handkerchief or cloth very loosely BETWEEN!!
the wound and the heart; put a stick, bayonet, or ramrod
between the skin and the handkerchief, and twist it around
until the bleeding ceases, and keep it thus until the surgeon
arrives.

If the blood flows in a slow, regular stream, a vein has been
pierced, and the handkerchief must be on the other side of
the wound from the heart, that is, *below* the wound.

A bullet through the abdomen (belly or stomach) is more
certainly fatal than if aimed at the head or heart; for in the
latter cases the ball is often glanced off by the bone, or follows round it under the skin; but when it enters the stomach
or bowels, from any direction, death is inevitable under all
conceivable circumstances, but is scarcely ever instantaneous.
Generally the person lives a day or two with perfect clearness
of intellect, often *not* suffering greatly. The practical bearing
of this statement in reference to the great future is clear.

Let the whole beard grow, but not longer than some three
inches. This strengthens and thickens its growth, and thus
makes a more perfect protection for the lungs against dust,
and of the throat against winds and cold in winter, while in
the summer a greater perspiration of the skin is induced, with
an increase of evaporation; hence, greater coolness of the parts
on the outside, while the throat is less feverish, thirsty and
dry.

Avoid fats and fat meats in summer and in all warm days.

Whenever possible, take a plunge into any lake or running
stream every morning, as soon as you get up; if none at hand,
endeavor to wash the body all over as soon as you leave your
bed, for personal cleanliness acts like a charm against all

diseases, always either warding them off altogether, or greatly mitigating their severity and shortening their duration.

Keep the hair of the head closely cut, say within an inch and a half of the scalp in every part, repeated on the first of each month, and wash the whole scalp plentifully in cold water every morning.

Wear woolen stockings and moderately loose shoes, keeping the toe and finger-nails always cut close.

It is more important to wash the feet well every night, than to wash the face and hands of mornings ; because it aids to keep the skin and nails soft, and to prevent chafings, blisters and corns, all of which greatly interfere with a soldier's duty.

The most universally safe position, after all stunnings, hurts, and wounds, is that of being placed on the back, the head being elevated three or four inches only ; aiding more than any one thing else can do, to equalize and restore the proper circulation of the blood.

The more weary you are after a march or other work, the more easily will you take cold, if you remain still after it is over, unless the moment you cease motion, you throw a coat or blanket over your shoulders. This precaution should be taken in the warmest weather, especially if there is even a slight air stirring.

The greatest physical kindness you can show a severely-wounded comrade is first to place him on his back, and then run with all your might for some water to drink ; not a second ought to be lost. If no vessel is at hand, take your hat ; if no hat, off with your shirt, wring it out once, tie the arms in a knot, as also the lower end, thus making a bag, open at the neck only. A fleet person can convey a bucketful half a mile in this way. I've seen a dying man clutch at a single drop of water from the finger's end, with the voraciousness of a famished tiger.

If wet to the skin by rain or by swimming rivers, keep in motion until the clothes are dried, and no harm will result.

Whenever it is possible, do, by all means, when you have to use water for cooking or drinking from ponds or sluggish streams, boil it well, and when cool, shake it, or stir it, so that the oxygen of the air shall get to it, which greatly improves it for drinking. This boiling arrests the process of fermentation which arises from the presence of organic and inorganic impurities, thus tending to prevent cholera and all bowel diseases. If there is no time for boiling, at least strain it through a cloth, even if you have to use a shirt or trowser-leg.

Twelve men are hit in battle, dressed in red, where there are only five dressed in a bluish gray, a difference of more than two to one; green, seven; brown, six.

Water can be made almost ice cool in the hottest weather, by closely enveloping a filled canteen, or other vessel, with woolen cloth kept plentifully wetted and exposed.

While on a march, lie down the moment you halt for a rest; every minute spent in that position refreshes more than five minutes standing or loitering about.

A daily evacuation of the bowels is indispensable to bodily health, vigor and endurance; this is promoted in many cases, by stirring a table-spoonful of corn (Indian) meal in a glass of water, and drinking it on rising in the morning.

Loose bowels, namely, acting more than once a day, with a feeling of debility afterward, is the first step toward cholera; the best remedy is instant and perfect quietude of body, eating nothing but boiled rice with or without boiled milk; in more decided cases, a woolen flannel, with two thicknesses in front, should be bound tightly around the abdomen, especially if marching is a necessity.

To have " been to the wars " is a life-long honor, increasing with advancing years, while to have died in defense of your country will be the boast and the glory of your children's children.

Dr. Phillips, late Army surgeon in the Crimea gives the following excellent advice in regard to

The Feet.

First, it is of the utmost importance that the men's boots should fit well. They will necessarily have a great deal of marching on a hot and perhaps sandy soil, more indeed than the English soldiers had, who were confined in a space of ten miles by about twenty, comprising the lower part of the Crimean Peninsula, from Balaklava to the Tchernaya River, and from Sevastopol to Baidar. While in charge of the fourth Division of the military train at the village of Kadikeid I had.

in two months, more than thirty soldiers entirely disabled from badly-fitting boots, out of a force of two hundred and thirty five English comprising the division, (not enumerating three hundred Turks, who often went barefooted, and whose feet frequently dropped off from frost-bites.)

These boots were generally too large, so that in walking they caused friction on the heels and produced ulcers. Such cases were not numerous among the officers, although several were similarly afflicted.

The British Government sent out boots of only two sizes, large and small, so that men with medium feet could not be fitted, and were obliged to take those or none. The men were in hospital, disabled from sore heels alone, being otherwise perfectly healthy. This may appear to many in civil life a mere trifle, but it was of serious import, as their services were entirely lost for many months. I may here remark, it is a most difficult matter to cure an ulcerated heel, on account of the skin being thick and sparingly supplied with blood vessels. The remedy I adopted in many cases (for the cure of the boots)—for many men applied for relief whom I did not consider sufficiently disabled to be admitted into hospital —was to cut a lozenge-shaped piece out of the boot over the instep, make two or three holes on either side, and order the men to lace them. This prevented the friction to a certain extent.

I would also recommend woolen stockings as being greatly superior to cotton, for men on the march. I have frequently found the smearing the feet with a tallow candle before putting on the stockings an admirable preventive of blistering.

I may also mention that it is a favorite habit of soldiers when they halt on the banks of a river, to take off their boots and wash their feet, and walk about barefooted while smoking their pipes; *nothing blisters the feet sooner than this practice.* When they are allowed to wash, the sergeants and corporals should insist on the feet being well dried, and the boots immediately replaced. This is a matter of great importance. Soldiers require as much care as children.

Dysentery.

In case of an attack of dysentery the soldier should get his hospital permit as soon as possible. Any ordinary disorder of the bowels may be corrected by a proper regard to diet and rest from duty. It is very aggravating to a diarrhea to be on the feet much, either at night or in the sun; hence, when the bowels are much loosened, the sergeant should report the man unfit for duty, and allow him all requisite aids to a cure. Rest and an abstinence from the usual ration food will, in most cases, effect a cure, and keep the soldier from the hospital. But diarrhea, aggravated by active duty, change of water and exciting diet, will soon become dysentery—one of the most fatal afflictions of camp-life. To avoid it the soldier should pay particular attention to his diet and his habits. Water had better be *boiled* before drinking if that is possible; if not, then decoct it with tea, or cut its edge with spirits. Weak soups, made of strained flour, or barley water, or of rice are to be resorted to—leaving off the usual salt rations. Go to bed early, sleep warmly in a *dry* spot. This course will avert many an attack which inevitably will result in a chronic dysentery.

A good authority before us says: "The first, the most important, and the most indispensable item in the arrest and cure of looseness of the bowels, is absolute quietude on a bed. Nature herself always prompts this by disinclining us to locomotion. The next thing is to eat nothing but common rice parched like coffee, and then boiled, and taken with a little salt and butter. Drink little or no liquid of any kind. Bits of ice may be eaten and swallowed at will. Every step taken in diarrhea, every spoonful of liquid, only aggravates the disease. If locomotion is compulsory, the misfortune of the necessity may be lessened by having a stout piece of woolen flannel bound tightly around the abdomen, so as to be doubled in front, and kept well in its place. In the practice of many years we have never failed to notice a gratifying result to follow these observances."

Procure the ingredients of the following of the surgeon and use it as directed: Take equal parts of syrup of rhubarb, paregoric, and spirits of campor; mix together. Dose for an adult, one tea-spoonful. If necessary, it may be repeated in two or three hours. This with quiet will surely restore the patient. The food to be used should be extremely non-irritant. We give a few recipes of such dishes as it is proper to use in cases of illness of any kind where light nourishment is required :

PANADA.—Having pared off the crust, boil some slices of bread in a quart of water for about five minutes. Then take out the bread, and beat it smooth in a deep dish, mixing in a little of the water it has boiled in ; and mix it with a bit of fresh butter, and sugar, and nutmeg to your taste.

TOAST-WATER.—Take a thin slice of stale bread, toast it brown on both sides slowly and equally. Lay it in a bowl, and pour on boiling water, and cover with a saucer to cool.

BEEF-TEA.—Take one pound of lean fresh beef cut thin, put it in a jar or wide-mouthed bottle, add a little salt, place it in a kettle of boiling water to remain one hour, then strain it, and there will be a gill of pure nourishing liquid. Begin with a tea-spoonful and increase as the stomach will bear. This has been retained on the stomach when nothing else could be, and has raised the patient when other means have failed.

CHICKEN, BEEF, OR VEAL BROTH.—This is made by cutting up the chicken, or the lean of veal or beef, and putting in two spoonfuls of washed rice, and boiling until tender. It may be used, if needed in haste, after boiling in less water about fifteen minutes, then filling it up and finishing. It should be put by in a bowl or pitcher covered, to keep for use. Warm it, and add crumbs of Boston crackers or bread a day or two old, with a little salt, and there is nothing more palatable for the sick.

WATER GRUEL.—Mix two table-spoonfuls of Indian or oatmeal with three of water. Have ready a pint and a half of boiling water in a saucepan or skillet, perfectly clean; pour this by degrees into the mixture in the bowl; then return it back into the skillet, and place it on the fire to boil. Stir it, and let it boil nearly half an hour. Skim it, and season it with a little salt. If it is admissible, a little sugar and nutmeg renders it more palatable. Also, if milk is not forbidden, a small tea-cupful added to a pint of gruel, and boiled up once, makes a nice dish for an invalid.

RICE GRUEL.—Take one spoonful of rice, a pint and a half of water, a stick of cinnamon or lemon-peel, boil it soft, and add a pint of new milk; strain it, and season with a little salt. If you make it of rice flour, mix one spoonful with a little cold water smoothly, and stir it into a quart of boiling water. Let it boil five or six minutes, stirring it constantly. Season it with salt, nutmeg, and sugar, and if admissible, a little butter. If the patient bears stimulants, a little wine may be added.

MILK PORRIDGE.—This is made nearly in the same way as gruel, only using half flour, and half meal, and half milk, instead of water. It should be cooked before the milk is added, and only boiled up once afterward.

MUTTON CUSTARD FOR BOWEL COMPLAINTS.—Take two ounces of fresh mutton suet shred fine, and a half dram of cinnamon, or some grated nutmeg, and boil in rather more than a pint of milk; when boiled, to be set by the fire till the scum rises, which should then be carefully taken off. Half a tea-cupful may be given warm or cold, as the patient prefers, three or four times a day. It should be continued till the complaint is quite cured.

FOR BREAD JELLY.—Measure a quart of boiling water, and set it away to get cold. Take one-third of an ordinary baker's loaf, slice it, pare off the crust, and toast the bread nicely to a light brown. Then put it into boiling water, set it on hot coals in a covered pan, and boil it gently, till you find, by putting some into a spoon to cool, that the liquid has become a jelly. Strain it through a thin cloth, and set it away for use. When it is to be taken, warm a tea-cupful, sweeten it with sugar, and add a little grated lemon-peel.   *

WINE WHEY.—Take half a pint of new milk, put it on the fire and the moment it boils, pour in that instant two glasses of wine and a tea-spoonful of powdered sugar previously mixed. The curd will soon form, and after it is boiled, set it aside until the curd settles. Pour the whey off and add a pint of boiling water, and loaf-sugar to sweeten to the taste. This may be drank in typhus and other fevers, debility, etc.

CALVES' FEET BROTH.—Boil two feet in three quarts of water until the water is half gone. Take off all the fat, season with a little salt, and, if suitable, a spoonful of white or port wine to a tea-cupful. This is nourishing and strengthening for an invalid. If a richer broth may be used, boil with the feet two ounces of veal or beef, a slice of bread, a blade or two of mace.

RICE JELLY.—Having picked and washed a quarter of a pound of rice, mix it with half a pound of loaf sugar, and just sufficient water to cover it. Boil it till it becomes a glutinous mass; then strain it; season it with whatever may be thought proper; and let it stand to cool.

HOT LEMONADE.—Cut up the whole of a lemon, rind and all, add one tea-cupful of white sugar, and pour on boiling water. This is good for colds, and is a pleasant drink for the sick.

These recipes are of most delicious dishes for the sick or convalescent, and can be made for the soldier in any ordinary hospital or camp, with a little exertion. We are indebted to Mrs. Victor's "Recipe Book" for them.

Fevers.

Fevers require hospital treatment. Only a physician should attempt to deal with them. Our advice to every soldier is to gain his hospital permit as soon as he finds an attack of fever impending.

Excellent Recipes for Various Cases

We here subjoin some excellent and very available recipes for various troubles which flesh is heir to in camp and campaign life. The soldier should preserve them and seek to aid his fellows when possible by their use upon others as well as upon himself:

THIEVES' VINEGAR.—Take of rue, sage, mint, rosemary, wormwood, and lavender, a large handful of each; infuse in one gallon of vinegar, in a stone jar closely covered, and keep warm by the fire for four days; then strain, and add one ounce of camphor, pounded; bottle, and keep well corked. There is a legend connected with this preparation (called in French *Vinaigre a quatre Voleurs*), that during the plague at Marseilles certain robbers plundered the infected houses with impunity, and being apprehended and condemned to death, were pardoned on condition of disclosing the secret of their preventive, as above. The mode of using is to wash the face and hands with it previous to exposure to any infection. It is very aromatic and refreshing in a sick-room, if nothing more.

PROTECTION AGAINST MUSKETO BITES.—Mix oil of pennyroyal with olive oil, and anoint the exposed parts of the person with it, when few if any insects will annoy one thus guarded. It is said that flies will not bite a horse if he is wet each morning with a decoction of walnut leaves.

FOR SPRAINS AND BRUISES.—Take one pint of train-oil, half a pound of stone-pitch, half a pound of resin, half a pound of bees-wax, and half a pound of stale tallow, or in like proportion. Boil them together for about half an hour, skim off the scum, and pour the liquid into cups, and when cold, it will be ready for use. When needed, it must be spread *as thick, but not thicker* than blister salve, upon a piece of coarse flaxen cloth. Apply it to the part sprained or bruised, and let it remain for a day or more; it will give almost immediate relief, and one or two plasters will be sufficient for a perfect cure.

ANOTHER.—In the Paris hospitals a treatment is practiced that is found most successful for a frequent accident, and which can be applied by the most inexperienced. If the ankle is sprained, for instance, let the operator hold the foot in his hands, with the thumbs meeting on the swollen part. These having been previously greased, are pressed successively with increasing force on the injured and painful spot for about a quarter of an hour. This application being repeated several times, will, in the course of the day, enable a patient to walk, when other means would have failed to relieve him.

RELIEF FOR A SPRAINED ANKLE.—Wash the ankle frequently with cold salt and water, which is far better than warm vinegar or decoctions of herbs. Keep your foot as cold as possible to prevent inflammation, and set with it elevated on a cushion. Live on very low diet, and take every day some cooling medicine. By obeying these directions only, a sprained ankle has been cured in a few days.

BRUISES, STINGS, ETC.—1. *For a bruise, etc.*—Bathe the part well with warm water, and afterward apply treacle spread on paper or linen, as most convenient; it soon heals, and no mark will be left. Treacle, if applied also in the early stages of a quinsy or sore throat, will speedily effect a cure. 2. *For the sting of a wasp or bee.*—Take about a wine-glassful of vinegar, put a little common soda into it, and bathe the parts affected. It gives almost immediate ease, and no pain or swelling will afterward be felt.

FOR A STING.—Bind on the place a thick plaster of common salt or saleratus moistened—it will soon extract the venom.

REMEDY FOR BLISTERED FEET FROM LONG WALKING.—Rub the feet, at going to bed, with spirits mixed with tallow dropped from a lighted candle into the palm of the hand.

DIRT IN THE EYE.—Place your forefinger upon the cheek-bone, having the patient before you; then draw up the finger, and you will probably be able to remove the dirt; but, if this will not enable you to get at it, repeat this operation while you have a netting-needle or bodkin placed over the eyelid; this will turn it inside out, and enable you to remove the sand or eyelash, etc., with the corner of a fine silk handkerchief. As soon as the substance is removed, bathe the eye with cold water, and exclude the light for a day. If the inflammation is severe, take a purgative, and use a refrigerant lotion.

POULTICE FOR A FESTER.—Boil bread in lees of strong beer; apply the poultice in the general manner. This has saved many a limb from amputation.

FROSTED FEET.—For frosted feet, deer's marrow will be found excellent. For chilblains, tincture of iodine; also muriatic acid, frequently applied, will relieve them.

FOR FROSTED FLESH.—Take chrome yellow and hog's lard, and make it into an ointment, and apply to the injured parts, warming the same into the skin.

COUGH MIXTURE.—Take one tea-cupful of molasses; add two table-spoonfuls of vinegar; simmer this over the fire; then, when taken off, add three tea-spoonfuls of paregoric, and as much refined niter as can be put upon the point of a small breakfast knife. Of this mixture, take two or three tea-spoonfuls on going to bed, and one or two during the day when you have a disposition to cough.

BLEEDING AT THE NOSE.—In obstinate cases, blow a little gum Arabic powder up the nostrils through a quill, which will immediately stop the discharge. Powdered alum is also good.

CERTAIN CURE FOR HEADACHE AND ALL NEURALGIC PAINS.—To be applied as any other lotion; opodeldoc, spirits of wine, sal ammoniac, equal parts.

TO STOP THE BLEEDING OF A WOUND.—Lay on the orifice, lint; if that is not sufficient, put on flour and then lint.

TO PREVENT WOUNDS FROM MORTIFYING.—Sprinkle sugar on them. The Turks wash fresh wounds with wine, and sprinkle sugar on them. Obstinate ulcers may be cured with sugar dissolved in a strong decoction of walnut leaves.

WARTS.—Wet them with tobacco juice, and rub them with chalk. Another: Rub them with fresh beef every day until they begin to disappear. This last is simple and effectual.

CORNS.—Take half an ounce of verdigris, two ounces of bees-wax, two ounces of ammonia; melt the two last ingredients together, and just before they are cold, add the verdigris. Spread it on small pieces of linen, and apply it, after paring the corn. This has cured inveterate corns.

BUNIONS.—Bunions may be checked in their early development by binding the joint with adhesive plaster, and keeping it on as long as any uneasiness is felt. The bandaging should be perfect, and it might be well to extend it round the foot. An inflamed bunion should be poulticed, and larger shoes be worn. Iodine twelve grains, lard or spermaceti ointment half an ounce, make a capital ointment for bunions. It should be rubbed on gently twice or thrice a day. Enlarged joints should be rubbed thrice a day with common salad oil, care being taken at the same time, not to strain or overtax the feet by too great or too frequent exercise. Slippers, and loose ones, should invariably be worn. On no account have tight-fitting shoes, slippers, or boots.

BOILS.—Make a plaster of molasses and flour, or honey and flour, and apply it as often as they get dry. If very painful, make a soft poultice of bread and milk, moistened with volatile liniment and laudanum. This will ease pain, allay inflammation, and hasten a cure. Remedies for cleansing the blood should be freely used.

To Remove Proud Flesh.—Pulverize loaf-sugar very fine, and apply it to the part affected. This is a new and easy remedy, and is said to remove it entirely without pain. It has been practiced in England for years.

Soyer's Hospital Recipes.

From Soyer's Crimean hospital experience we have obtained many good suggestions for the benefit of troops. We may here add to our already liberal quotations such as seem to us particularly desirable and useful :

Boiled Rice, semi-curried, for the premonitory symptoms of Diarrhea.—Put one quart of water in a pot or saucepan; when boiling, wash one and a half pound of rice and throw it into the water; boil fast for ten minutes; drain your rice in a colander, put it back in the saucepan, which you have slightly greased with butter, and let soak all night; and prior to using it, wash it and squeeze with your hands, to extract the salt. In case the meat is still too salt, boil it for twenty minutes, throw away the water, and put fresh to your stew. By closely following the above receipt you will have an excellent dish.

Plain Oatmeal.—Put in a pan one-quarter of a pound of oatmeal, one and a half ounce of sugar, half a tea-spoonful of salt, and three pints of water; boil slowly for twenty minutes, " stirring continually," and serve. A quarter of a pint of boiled milk, an ounce of butter, and a little pounded cinnamon or spice added previous to serving is a good variation. This receipt has been found most useful at the commencement of dysentery by the medical authorities.

Beef Tea—Receipt for six Pints.—Cut three pounds of beef into pieces the size of walnuts, and chop up the bones, if any; put it into a convenient-sized kettle, with one and a half pound of mixed vegetables, such as onions, leeks, celery, turnips, carrots (or one or two of these, if all are not to be obtained), one ounce of salt, a little pepper, one tea-spoonful of sugar, two ounces of butter, half a pint of water; set it on a sharp fire for ten minutes or a quarter of an hour, stirring now and then with a spoon, till it forms a rather thick gravy at bottom, but not brown; then add seven pints of hot or cold water, but hot is preferable; when boiling, let it simmer gently for an hour; skim off all the fat, strain it through a sieve, and serve.

Essence of Beef Tea—*For Camp Hospitals.*—" Quarter pound tin case of essence." If in winter, set it near the fire to melt; pour the contents in a stewpan and twelve times the case full of water over it hot or cold; add to it two or three slices of onion, a sprig or two of parsley, a leaf or two of celery, if handy, two tea·spoonfuls of salt, one of sugar; pass through a colander, and serve. If required stronger, eight cases of water will suffice, decreasing the seasoning in proportion. In case you have no vegetables, sugar or pepper, salt alone will do, but the broth will not be so succulent.

THICK BEEF TEA.—Dissolve a good tea-spoonful of arrow-root in a gill of water, and pour it into the beef tea twenty minutes before passing through the sieve ; it is then ready.

STRENGTHENING BEEF TEA, WITH CALVES' FOOT JELLY OR ISINGLASS.—Add one-quarter calves' foot gelatine to the above quantity of beef tea previous to serving, when cooking.

MUTTON AND VEAL TEA.—Mutton and veal will' make good tea by proceeding precisely the same as above. The addition of a little aromatic herbs is always desirable. If no fresh vegetables are at hand, use two ounces of mixed preserved vegetables to any of the above receipts.

CHICKEN BROTH.—Put in a stewpan a fowl, three pints of water, two tea-spoonfuls of rice, one tea-spoonful of salt, a middled-sized onion, or two ounces of mixed vegetables; boil the whole gently for three quarters of an hour; if an old fowl, simmer from one hour and a half to two hours, adding one pint more water; skim off the fat, and serve. A small fowl will do.

Note—A light mutton broth may be made precisely the same by using a pound and a half of scrag of mutton instead of fowl. For thick mutton broth proceed as for thick beef tea, omitting the rice; a table-spoonful of burnt sugar-water will give a rich color to the broth.

PLAIN BOILED RICE.—Put two quarts of water in a stewpan, with a tea-spoonful of salt; when boiling, add to it half a pound of rice, well washed; boil for ten minutes, or till each grain becomes rather soft; drain it into a colander, slightly grease the pot with butter, and put the rice back into it; let it swell slowly for about twenty minutes near the fire, or in a slow oven; each grain will then swell up, and be well separated; it is then ready for use.

SWEET RICE.—Add to the plain boiled rice one ounce of butter, two table-spoonfuls of sugar, a little cinnamon, a quarter of a pint of milk; stir it with a fork, and serve; a little currant jelly or jam may be added to the rice.

RICE WITH GRAVY.—Add to the rice four table-spoonfuls of the essence of beef, a little butter, if fresh, half a tea-spoonful of salt; stir together with a fork, and serve.

SAGO JELLY.—Put into a pan three ounces of sago, one and a half ounce of sugar, half a lemon-peel cut very thin, one-quarter tea-spoonful of ground cinnamon, or a small stick of the same; put to it three pints of water, and a little salt; boil ten minutes or rather longer, stirring continually until rather thick, then add a little port sherry, or Marsala wine; mix well, and serve hot or cold.

ARROW-ROOT MILK.—Put into a pan four ounces of arrow-root, three ounces of sugar, the peel of half a lemon, one-quarter tea-spoonful of salt, two and a half pints of milk ; set it on the fire, stir round gently, boil for ten minutes, and serve. If no lemons at hand, a little essence of any kind will do. When short of milk, use half water ; half an ounce of fresh butter is an improvement before serving ; if required thicker, put in a little milk.

THICK ARROW-ROOT PANADA.—Put in a pan five ounces of arrow-root, two and a half ounces of white sugar, the peel of half a lemon, a quarter of a tea-spoonful of salt, four pints of water, mixed well ; set on the fire, and boil for ten minutes ; it is then ready. The juice of a lemon is an improvement ; a gill of wine may also be introduced, and one ounce of calves' foot gelatine previously dissolved in water, will be strengthening. Milk, however, is preferable, if at hand.

ARROW-ROOT WATER.—Put into a pan three ounces of arrow-root, two ounces of white sugar, the peel of a lemon, one-half tea-spoonful of salt, four pints of water ; mix well, set on the fire, and boil for ten minutes. It is then ready to serve either hot or cold.

RICE WATER.—Put seven pints of water to boil, add to it two ounces of rice, washed, two ounces of sugar, the peel of two-thirds of a lemon ; boil gently for three-quarters of an hour ; it will reduce to five pints ; strain through a colander ; it is then ready. The rice may be left in the beverage, or made into a pudding, or by the addition of a little sugar or jam, will be found very good for either children or invalids.

BARLEY WATER.—Put in a saucepan seven pints of water, two ounces of barley, which stir now and then while boiling ; add two ounces of white sugar, the rind of half a lemon, thinly peeled ; let it boil gently for two hours without covering it ; pass it through a sieve or colander ; it is then ready. The barley and lemon may be left in it.

SOYER'S PLAIN LEMONADE.—Thinly peel the third part of a lemon, which put into a basin with two table-spoonfuls of sugar ; roll the lemon with your hand upon the table to soften it ; cut it into two, length-wise, squeeze the juice over the peel, etc., stir round for a minute with a spoon to form a sort of syrup ; pour over a pint of water, mix well, and remove the pips ; it is then ready for use. If a very large lemon, and full of juice, and very fresh, you may make a pint and a half to a quart, adding sugar and peel in proportion to the increase of water. The juice only of the lemon and sugar will make lemonade, but will then be deprived of the aroma which the rind contains, the said rind being generally thrown away.

THE LAW OF PRIZES.

PRIZE, from the French *prise*, is the taking at sea of a vessel by a belligerent power with intent of appropriating to the use of the captor the ship, or cargo, or both. The subject of the capture is also called a prize. In order to make a valid title to the prize, a trial must be had before a court of competent jurisdiction to ascertain the true character of the prize, and a sentence or degree of condemnation must be passed or made in due form. The claimant of the ship and cargo or of either has a right to appear and be heard in defense of his claim, and witnesses are examined either orally or by commission, according to the usual practice of courts. A belligerent has a right to take his prize into any port of a neutral power, but no prize court of a belligerent can sit in a neutral country, but it must sit either in the country of the captor, or else in the country of an ally. The prize, however, may remain in a neutral port while being adjudicated upon by the prize court of the captor or captor's ally, and a good title is made by the sentence of condemnation, although the proceeding is in legal language, *in rem*, and the subject or *corpus* out of the jurisdiction of the court. It is to be understood that neutral powers may refuse to belligerents the right to bring prizes into their ports, unless the right is guaranteed by treaty.

Ships and cargoes belonging to neutrals are likewise subject to capture and confiscation by belligerents for various offenses. First, for attempting to violate a blockade after reasonable notice of its existence. Even sailing for a blockaded port, or standing off and on, subjects neutral ships and cargoes to condemnation, whether the port be reached or not. But the blockade must in all cases be an actual, and not a constructive or paper blockade; otherwise no rightful sentence of condemnation can be passed. Secondly, neutral ships and cargoes are subject to capture and condemnation for carrying to an enemy's country articles contraband of war, such as arms, ammunition, and naval stores, and also provisions when carried to an enemy's navy or a place besieged, or for carrying dispatches or soldiers to the enemy. Thirdly, neutral ships and cargoes may be captured and condemned for resisting a belligerent's right of search. Fourthly, such ships and cargoes may be lawfully captured and condemned for sailing under the enemy's flag, or with his pass or license. Fifthly, it is held by English courts that a neutral engaging in the enemy's coasting trade is subject to capture and condemnation, but our own courts have doubted this doctrine. These general principles are subject to many distinctions of a nice .character.

By the law of nations, all produce of a hostile soil found in a neutral's ship may be seized in transitu. Most of our treaties, however, provide that free ships make free goods, and it is probable that, even in the absence of treaties, and supposing that we are not bound by the Treaty of Paris (1859), our Government would always recognize this principle.

A COMPLETE DICTIONARY

OF

MILITARY TERMS AND SCIENCE.

Abattis. A kind of outer intrenchment, consisting of young trees, felled and laid upon the ground a short distance from the parapets of field works, with the points of their larger branches sharpened and extending outward, for retarding the enemy's advance.

Absence, with Leave and without Leave. Officers, non-commissioned officers and privates are said to be *absent with leave*, when they have obtained permission to that effect; *absent without leave*, when they fail to join their regiments on the expiration of their leave.

Accouterments. The belts, pouches, cartridge boxes, etc., of the soldier. The belts, sashes, etc., of the officers, are termed *appointments*, by the British authorities.

Acquittance-Roll. A roll containing the names and signatures of the privates of each troop or company of a regiment, and showing their respective debits and credits.

Adjutant. The assistant of a commanding officer of a regiment, in the details of regimental duty and discipline.

Adjutant-General. The officer who assists the general of an army in the general details of his duties.

Adjutant-General of the Forces. The chief officer of the general staff.

Advance Guard. The detachment of troops which precedes the march of the main body. *The Rear Guard* is that which covers its rear.

Agent (Army). The person who transacts the pecuniary business of regiments.

Aid-de-Camp. An officer on the personal staff of a general or field officer, to receive and distribute his orders.

Aiguillette. A decoration consisting of tagged points of bullion cord and loops, worn on the right shoulder of officers of the cavalry.

Aim (To Take). To mark out the object to be struck by a cannon or musket ball.

Alarm Post (in the Field). Is the ground appointed by the quartermaster-general for each regiment or detachment, to march to in case of alarm. In a garrison, it is the place allotted by the governor for the troops to draw up in, on any sudden alarm.

Alignment. A formation in straight lines. *The alignment of a battalion* is the position of a body of troops drawn up in line. The alignment of a camp signifies the relative position of the tents, etc., so as to form a straight line from given points.

Altimetry. The taking or measuring of altitudes or heights.

Altitude. In cosmography, is the perpendicular height of an object, or its distance from the horizon upward. Altitudes are *accessible* and *inaccessible.* *Accessible altitude* of an object is, that to whose base access can be had, to measure the distance between a given point and the foot of the object on the ground. *Inaccessible altitude* of an object is, that when the foot or bottom of the object can not be approached, on account of some obstacle, as water, etc. *Altitude of a shot or shell* is the perpendicular height of the vertex of the curve in which it moves above the horizon.

Ambulant. Changing positions according to circumstances. An ambulant hospital is that which follows an army.

Ambuscade. A detachment of troops placed in concealment to surprise or attack an enemy.

Ambush. A place of concealment from which an enemy may be surprised by a sudden attack.

Amende Honorable. Satisfaction for an offense committed against the rules of honor or military etiquette.

Amplitude. In gunnery, the range of shot, or the horizontal line which measures the distance it has reached.

Ammunition. Powder and ball, shells, bullets, cartridges, grapeshot, tin and case-shot, carcasses, grenades, etc. The ammunition for firearms is *fixed* and *unfixed.* The *fixed* comprises loaded shells, carcasses, and cartridges, filled with powder; also shot fixed to powder. Ball and blank cartridges are also termed *fixed* ammunition. *Unfixed* ammunition is round, case, and grape-shot, or shells, not filled with powder.

Approaches. Sunk works and passages carried on toward a besieged fortress, as the trenches, saps, mines, etc.

Armistice. A truce or temporary suspension of hostilities.

Arms (Bells of) or *Bell Tents.* Tents in the shape of a cone, in which each company's arms are piled in the field.

Arms (Stand of). A complete set of arms for one soldier.

Army. Armies are (1) A covering army; (2) A blockading army; (3) An army of observation; (4) An army of reserve; (5) A flying army; (6) The grand and main army; (7) The "standing army," which, in the United States, is always limited by special acts of Congress.

Articles of War. Rules and regulations for the government of the army.

Arsenal. The place where warlike instruments of all kinds are deposited.

Artillery. All projectile machines of war, as cannons, mortars, howitzers, etc., with the requisite apparatus and stores for field and siege service. *A Train of Artillery* consists of the attendants and carriages which accompany the artillery into the field. *A Park of Artillery* is the place where the artillery ammunition is encamped ready for service. It is also used to imply a heavy complement of guns.

Back Step. The retrogade movement of a man or body of men without changing front..

Backward. The retrogade movement of troops from line into column, and *vice versa.*

Baggage. The clothes, tents, utensils, etc., belonging to a regiment or an army.

Bags. In military operations, are either sand bags or earth bags; and are used either to repair breaches and the damaged embrasures of batteries, or to raise a parapet in haste, or to repair one that is beaten down by the enemy's fire.

Balls or *Bullets.* Consist of lead or iron for the use of small arms or artillery; or they are light or fire balls, and smoke balls. The light or fire ball, which is spherical or oblong, is used during sieges, for the purpose of discovering parties at work, etc. The smoke balls are thrown from mortars to annoy besiegers, continuing to smoke for about half an hour.

Band. The body of musicians attached to every regiment or battalion.

Banderols. Small flags used in marking out a camp, etc.

Banquette. An elevation or step constructed along the interior of the parapet, to enable the shortest men to fire over it.

Barbette. A platform raised behind a parapet or breast-work, that the guns mounted upon it may have a free range over the surrounding country. Guns so placed are said to be mounted *en barbette.*

Barricades. Obstructions formed in streets and highways, consisting of abattis, breast-works, overturned wagons, carts, etc., to prevent an enemy's access.

Barriers. Pointed stakes to prevent cavalry or infantry from suddenly rushing in on the besieged.

Base Line, or *Base of Operations.* The frontier or line of fortresses, on which all the magazines and means of supply of an army are established, and from which the lines of operation proceed.

Bastion. A projection or salient angle from the general outline of a fortress, with an opening toward the body of the place called a gorge.

Baton. The staff or truncheon which is the symbol of a field-marshal's authority.

Battalion. A body of infantry of two or more companies.

Battalion Men. The soldiers, except those of the two flank companies, belonging to the different companies of an infantry regiment.

Battering Train. A train of artillery used solely for besieging fortresses, *inclusive* of mortars and howitzers.

Battery. A number of pieces of ordnance, consisting either of guns, howitzers, or mortars, according to the service for which they are required.

Battle Array, or *Line of Battle.* The order or arrangement of troops in battle.

Battlements. Notches or indentures in the top of old castles or fortified walls, or other buildings, in the form of embrasures, for the greater convenience of firing or looking through.

Bayonet. A weapon first used by the French in 1671, and deriving its name from having been first manufactured at Bayonne. It is now an instrument of war, constructed of several shapes, that is considered of invaluable utility to infantry. It alone is regarded equal to cope with a cavalry charge, while in a direct charge it is used with terrible effect by a well trained soldiery. It was the favorite weapon of Napoleon; and by it Garibaldi has achieved his most brilliant victories. The *sword bayonet*, recently introduced, is a truly formidable instrument of death. It can be made use of either as a cutting or thrusting instrument. By poising it horizontally, like a quarter-staff, as high as the head of his adversary, the soldier, by a slight movement in the segment of a circle, can sweep its sharp blade across the neck, face and breast of three men opposed to him in line.

Besiege (*To*). To invest a fortified town with an armed force.

Billeting. The quartering of troops in the houses of towns and villages.

Bivouac. Troops are said *to bivouac* when they do not encamp, but lie under arms for the night.

Blackhole. A place for confinement of soldiers guilty of insubordination or criminality—more generally called the "Guard House."

Block-house. A wooden fort.

Body of a Place. The main line of bastions and curtains, or the space inclosed by the *enceinte* of a fortress.

Boom. A cable or chain floated with masts or spars, placed across the mouth of a river or harbor, to bar the access of an enemy.

Break Ground (*To*). Commencing the siege of a fortress by opening the trenches.

Breach Loader. A piece which is loaded at the breach instead of at the muzzle. *A Muzzle Loader* is one which receives the charge at the muzzle.

Breast-work. A parapet breast-high.

Brevet-Rank. A rank in the army higher than that for which pay is received; and which gives precedence, according to the date of the commission, when corps are brigaded.*

Brevet (*The*). A term used to express promotion *by honor*.

Bridge. See *Pontoon* and *Pontooning.*

Bridge-Head. See *Tete du Pont.*

Brigade. A division of troops, consisting of two or more regiments, under command of a brigadier-general. *Mixed Brigade* is composed of infantry and cavalry, generally used as advance guards.

---

* In the celebrated controversy between Generals Scott and Gaines (1828–29) Congress decided that a brevet did not confer actual rank, and that the order of promotion must follow only the official commission,

Brigade-Major. An officer charged with the detail of the duties of a brigade.

Brown Bess. A sobriquet or nickname, for the old regulation (English) musket.

Bugle-Calls. The sounds of the bugle used in the field or on parade, where the voice would be ineffectual to convey commands.

Bugler. The person who sounds the bugle for advancing, skirmishing, or retreating maneuvers.

Bulletin. The official account of public transactions and military operations. It is more commonly called " Official Report."

Cadence. In tactics, a uniform time and pace in marching.

Cadets. Youths educated at the Military Academy at West Point at the expense of the United States Government. They graduate, after a term of five years, with the rank of second lieutenant in the United States army.

Caisson. An ammunition wagon or tumbril. Also a wooden frame or chest, containing loaded shells, and buried at the depth of five or six feet under some work the enemy appears desirous to possess, and which, when he has become master of, is fired by means of the train conveyed through a pipe to it, when the shells becoming inflamed, the assailants are blown up.

Calibre. In gunnery, the diameter *of the bore* of a cannon.

Caltrops, or *Crow's Feet.* Pieces of iron having four points, so disposed that three of them always rest upon the ground, and the fourth stands upward in a perpendicular direction. Each point is three or four inches long. They are scattered over the ground and passages where the enemy is expected to march, especially the cavalry, in order to embarrass their progress.

Camp. The extent of ground occupied by the tents of an army when in the field. For camp disposition of the United States army see " Army Regulations," pages 67–71.

Camp-Colors. The flags or ensigns which mark out the lines of an encampment. Also, small colors placed on the right and left of the parade of a regiment when in the field.

Camp-Color-Men. Those soldiers who carry camp-colors to the field on days of exercise, and plant them to mark out the lines.

Camp (Flying). A strong body of cavalry and infantry always in motion, to cover its own garrisons, and to keep the army of the enemy in continual alarm.

Cannon. Cannons were originally made of iron bars soldered together, and fortified with strong iron hoops. Others were made of thin sheets of iron rolled up together, and hooped ; and on emergencies they were made of *leather*, with plates of iron or copper ! For a descriptive of the cannon now in use in the American army, see article, page 64.

Canteen. A tin or wooden vessel, in which soldiers carry water or other liquid on the march. The term also signifies a suttling house kept in a garrison or barrack-yard, for the supplying of troops.

Cantonments. The situations in which troops are quartered in towns and villages.

Caponniére. A passage from the body of a place across the ditch to an outwork.

Carbineers or *Carabineers.* Horsemen armed with carabines, who occasionally act as infantry.

Carcasses. Shells containing a composition of combustibles projected from mortars.

Carronade. A short piece of iron ordnance, originally made on the Carron in Scotland.

Cartel. An agreement for a mutual exchange of prisoners.

Case or *Canister Shot.* Bullets, pieces of iron, etc., inclosed in a circular tin case, and discharged from heavy pieces of ordnance.

Casemate. A cave under the rampart, with loopholes through which artillery may be discharged.

Cashiered. Dismissed with ignominy from the service.

Castrametation. The planning and tracing out an encampment.

Casuals or *Casualties.* A term implying soldiers who die, desert, or have been discharged.

Cat-o-nine-Tails. A whip with five or nine knotted cords used for flogging military offenders. See Articles of War regarding its use, page 20.

Cavalry. Horse soldiers.

Chace of a Gun. Its entire length.

Chamade. See *Drum.*

Chamber of a Cannon. (or mortar). That part of the bore which receives the charge of powder. *Chamber of a Mine,* is the place where the charge of powder is lodged for the purpose of blowing up the works over it. *Chamber of a Battery*, is a place sunk under ground for holding powder, loaded shells and fusees, where they may be out of danger, and preserved from rain or moisture.

Cheeks of a Gun-carriage. The strong planks forming its sides.

Chandelier. A moveable parapet, consisting of wooden frames, filled with facines laid to cover working parties in the trenches.

Chevaux-de-Frize (*Friesland Horses*). Obstacles consisting of a beam of timber, with strong stakes pointed with iron, driven through it in different directions, used for defending avenues and passages, for impeding river channels, stopping up breaches, and impeding assaults. [The term takes its derivation from the apparatus having been first used at the siege of Groningen, in Friesland, in the year 1658, against the cavalry of the enemy].

Chevrons. The marks on the sleeves of the coats of non-commissioned officers.

Circumvallation (*Line of*). A fortification of earth, consisting of a parapet or breastwork and trench, to cover the besiegers against any attempt of the enemy, in favor of the besieged.

Club (*To*). In a military sense, to throw into confusion; to deform through ignorance or inadvertency. *To Club a Battalion* is to throw it into confusion. The more common use of the word, however, in this country, implies to use the musket or rifle as a club in a close fight.

Color-Sergeant. The regimental sergeant whose duty it is to attend to the colors in the field.

Colors (*Regimental*). Are two in each regiment, one the national ensign stars and stripes, the other the regimental color. [See "Army Regulations," 1368–69–70–71–72–73 for special regulations for artillery, cavalry, etc.]

Column. A body of troops in deep file and narrow front. Troops are *in close column* when they are close together; in *open column*, when there are intervals sufficient for wheeling into line when requisite.

Commissary. That department of military economy which is charged with the care of the provisions, tents, etc., of an army.

Contribution. An imposition or tax paid, in provisions or money, by the inhabitants of a town or country to an enemy.

Convention. An agreement for the suspension of hostilities, or the evacuation of a post, etc.

Cordon. A chain of posts, or an imaginary line of separation between two hostile armies, either in the field or in winter-quarters.
Also used to signify bodies of troops stationed at detached intervals filled up by unceasing patrolling, for preventing the escape
of an enemy, or to prevent his sudden irruption into a country.
Also used to express the coping of the escarp of the ditch of a
fortress.

Corporal (in the army). A non-commissioned officer under the sergeant. His duty is to place and relieve sentinels and to take
charge of a squad in drill. (In the navy), an inferior officer under
the master at arms.

Corporal (*Lance*). A soldier who acts as corporal, with only the
pay of a private.

Corps. A body of troops.

Corps d'Armée. A portion of a grand army possessed of all the con-
stituents of a separate or an independent army.

Cover. In military parlance, signifies security or protection.

Covered Way. A space extending from the counterscarp to the
crest of the glacis, and surrounding the body of the fortress with
its outworks.

Counter Approach. A trench or passage carried out by the besieged to counteract the works of the besiegers.

Counterforts or *Buttresses.* Solid works of masonry built behind
walls to strengthen them.

Counter-Guard. A work placed before bastions to cover the opposite flanks from being seen from the covered way.

Countermarch (*To*). To change the front of an army, battalion, etc.,
by an inversion of its several component parts.

Counter Mining. (See *Mining*).

Counterscarp. The exterior slope of the ditch of a fortress.

Countersign. A word or number exchanged between sentries on
duty in camp or garrison. Also the watchword demanded by
sentries from those who approach their posts.

Counter Trench. (See *Trench*).

Countervallation (*Line of*). A breastwork with a ditch before it to
defend the besiegers against the enterprises of the garrison.

Coup-de-Main. A sudden and vigorous attack.

Coup-d'-Œil. The seeing at a glance of the eye, the features of a
country, or the position of an enemy. The term also implies the

judicious selection of the most advantageous position for an encampment, or a field of battle.

Court of Inquiry. A meeting of officers to inquire into the conduct of a commander of an expedition, to ascertain whether there be ground for a court-martial.

Courts-Martial. Military courts appointed for the investigation and punishment of offenses committed by officers and soldiers in breach of the articles of war: they are three; 1. *General;* 2. *District;* and 3. *Regimental.* See "Army Regulations," 861, *et seriatim.*

Crémaillère. An indented or zigzag outline, resembling the teeth of a saw.

Crenellated. Loop-holed.

Crenaux. Loop-holes.

Culverin. A long cannon.

Cunette. A trench in the middle of a dry ditch.

Cuirass. Defensive armor, covering the body of the wearer from the neck to the waist. Not now used.

Cuirassiers. Heavy cavalry, clad in cuirasses. Not in service now.

Curtain. That part of the rampart which connects two contiguous bastions.

Cut-off. In military parlance, signifies to intercept, or hinder from union or return.

Cylinder of a Gun. The whole length of the bore of a piece of ordnance.

Debouch. The outlet of a wood or narrow pass. *To debouch,* is to march out of a defile, narrow pass, or wood. *Debouchment* is the marching out of a defile, etc., into open ground.

Debris. The wreck or remains of a routed army.

Decimation. The infliction of death on every tenth man of a corps; used also to imply great slaughter.

Decoy. A stratagem to carry off the horses of a foraging party of the enemy, or from pasture. Also implies any deception.

Defaulter's Book. A regimental record of the offenses and irregularities of the privates and non-commissioned officers of a regiment.

Defilement or *Defilading.* The arrangement of the plan and profile of the works of a fortress, so as to prevent their being enfiladed.

Defile. A narrow passage or road, through which troops can not march, otherwise than by making a small front, and filing off. *To Defile* is to move off in a line, or file by file. The term also implies the reduction of divisions and subdivisions of troops to a small front, to enable them to march through a defile.

Deliver Battle (*To*). Is when hostile armies are in sight of each other, to commence an attack.

Deploy (*To*). To display or spread out. A column is said *to deploy*, when the divisions open out or extend, for the purpose of making a flank march, or to form in line on any given division. *Deployment* is the act of unfolding or expanding a given body of troops, so as to extend their front.

Depôt. A place where military stores are deposited. The term also signifies the station of the reserve companies of regiments.

Depression. The pointing of a piece of ordnance so that the shot may be projected under the point blank line.

Depth of a Battalion or *Squadron.* The number of men in rank and file.

Detach (*To*). To send out a body of men on some particular service, distinct from that of the main body.

Detachment. A number of men drawn out from several regiments or companies.

Detonating Powder. That part of the cartridge which is detached for priming.

Diminish (*To*) *the Front of a Battalion.* To adapt the column of march or maneuver, according to the obstructions and difficulties which it meets in advancing.

Disengage (*To*) *a Column* or *Line.* To clear a column or line which may have lost its proper front by the overlapping of any particular division, company, or section, when ordered to form up.

Disengage (*To*) *the Wings of a Battalion.* When the battalion countermarches from its center, and on its center, by files.

Dislodge (*To*). To drive an enemy from his post.

Dismantle a Fortification (*To*). To render it incapable of defense.

Dismantle a Gun (*To*). To render it unfit for service.

Dismount (*To*) *Cannon.* To break their carriages, wheels, axle-trees, etc., so as to render them unfit for service.

Disobedience of Orders. Any infraction, by neglect or willful omission, of general or regimental orders.

Division of an Army. A body of troops, consisting of two or more brigades, under the command of a general of division.

Dock-Yard Battalions. A defensive force, consisting of the superintendents, clerks, and laborers of the respective navy or dock-yards.

Draw Up (To). To form in battle array.

Drawn Battle. A battle in which both sides claim the victory.

Dress (To). To arrange a company or a battalion in such a position or order, that an exact continuity of line is preserved in the whole front, or in whatever direction it is to be formed. *Dressing* is effected by each man taking short quick steps, until he gradually obtains his position in the rank or line.

Dressers. Those men who take up direct or relative points by which a corps is enabled to preserve a regular continuity of front, and to exhibit a straight alignment.

Drill (To). To teach recruits the first principles of military movements and positions, etc.

Drum (Beats of). The various beats of the drum are: *The General,* to give notice to the troops that they are to march: *The Assembly, The Troop,* to order the troops to repair to the place of rendezvous, or to their colors: *The March,* to command them to move, always with the left foot first: *Tat-too* or *Tap-too,* to order all to retire to their quarters: *To Arms!* for soldiers who are dispersed, to repair to them: *The Reveillé* always beats at break of day, to warn soldiers to rise, and sentinels to forbear challenging, and to give leave to come out of quarters: *The Retreat,* a signal to draw off from the enemy; also, a beat in both camp and garrison a little before sunset, when the gates are shut, and soldiers repair to their barracks: *The Alarm,* to give notice of sudden danger, that all may be in readiness for immediate duty: *The Parley, (Chamade)* a signal to demand some conference with the enemy: *The Sergeants Call,* a beat for calling the sergeants together in the orderly-room, or in camp to the head of the colors: *The Drummers' Call,* a beat to assemble the drummers at the head of the colors, or in quarters at the place where it is beaten: *The Preparative,* a signal to make ready for firing. As soon as it commences, the officers step out of the rank, and when it has ceased, the several firings commence. When *the General* is beat, they fall back into the front rank: *The Long Roll,* a signal for the assembly of troops at any parade. These beats are either *ordinary* or *extraordinary*.

Drum-Major. The instructor of the drummers in the beats.

Echelon. A formation of divisions of a regiment, or of entire regiments, resembling the steps of a ladder. Echelon movements and positions are not only necessary and applicable to the immediate attacks and retreats of large bodies of troops, but also to the previous oblique and direct changes of situation, which a battalion, or larger corps of troops already formed in line, may be compelled to make to the front or rear, or on a particular fixed division of the line.

Effective. Fit for service, in contradistinction to *non-effective*, or unfit for service.

Empilement. The act of disposing balls, grenades, and shells, in the most secure and convenient manner.

Embrasures. The openings which are made in the parapets of a work, for the purpose of pointing cannon against objects.

Encampment. The pitching of a camp.

Enceinte. The outline of a fortress, including the ramparts, wall, ditches, etc.

Enfilade (*To*). To sweep or rake the whole length of any work, or a line of troops, by a fire from a battery on the prolongation of that line.

Engarrison (*To*). To protect a place by a garrison.

Enrol (*To*). To enlist men to serve as soldiers.

Entrepôts. Magazines, as also places appropriated in garrison towns, for the reception of stores, etc.

Epaulement. A mound of earth thrown up to cover troops from the enemy's fire.

Eprouvette. A machine to prove the strength and quality of gunpowder.

Equalize (*To*) *a Battalion.* To tell off a certain number of companies in such a manner that the several component parts shall consist of the same number of men.

Equipage (*Camp* or *Field*). Tents, cooking utensils, saddle horses, bat-horses, baggage-wagons, etc.

Equipment. The complete dress of a soldier, including arms, accouterments, etc.

Escalade (*To*). To scale the walls of a fortress.

Escarp. The sides of the ditch next to the rampart.

Escort. Troops who guard prisoners on a march to prevent their escape.

Esplanade. An open space of ground separating the citadel from the town.

Estaffette. A military courier, sent express from one part of an army to another.

Evolutions. The changes of the position of troops, either for attack or defense.

Exempts. Persons exempted from certain services, or entitled to peculiar privileges.

Extraordinaries. Allowances for the expenses of barracks, marches, encampments, etc.

Faces of a Square. The four sides of a battalion when formed in square.

Facings. The different movements of a battalion, or any other body of troops, to the right, to the left, or to the right (*or* left) about, or to the right (*or* left) half face, or to the right (*or* left) three-quarter face. The term also implies the different facings the recruit is taught while at drill.

Fall (*To*) *Back.* To recede from a position recently occupied.

Fall (*To*) *In.* To form in ranks in parade, line, division, etc.

Fall (*To*) *Out.* To quit the rank or file.

False Attack. A feigned attack for the purpose of diverting the enemy from the real point of attack.

Fascines. Faggots made of brushwood or small branches of trees, of various dimensions, according to the purposes for which they are intended.

Feint. A false or mock attack or assault made for the purpose of concealing a real one. *A Feint*, when in an aggressive attitude, is threatening one part of an opponent's person, when it is intended to try his vulnerability on another part.

Feu de Joie. A discharge of musketry, or of salvos of artillery in celebration of some important event.

Field-Days. Days on which troops are taken out to the field, to be instructed in field exercise and evolutions.

Field-Officer. An officer above the rank of captain, and under that of general, namely; majors, lieutenant-colonels, and colonels.

File. A line of soldiers drawn up one behind another. The term also signifies two soldiers, the front and rear rank men. *To file* is to advance to or from a given point by files, and *to file off* or *defile*, is to wheel off by files. *File leader* is the front man of a battalion or company standing two deep. *File-marching* is when soldiers so follow one another that every man in the first rank appears to lead a file. *Filings* are movements to the front, rear, or flanks, by files.

Fire (Running). Is when a line of troops fire rapidly in succession, or one after another.

Flag of Truce. A flag carried when some pacific communication is to be made to the enemy. It is generally white colored.

Flank (To). Is to take up a position without being exposed to all the enemy's fire. To outflank, is the increasing the front of a body of troops, till it outstretches the opposing forces.

Flank en Potence. Is where the extremity of the right or left wing is thrown back at an obtuse angle in the rear of the line.

Forlorn Hope. The party or body of men and officers, who lead the storm of a fortress. In the French service, this devoted band is emphatically styled *les enfans perdus*.

Formation. The arrangement or drawing up of troops according to prescribed rules. Formations are at *close order* and *open order*. The term also signifies the constituent or component parts of a regiment. An infantry regiment consists of companies and battalions; a cavalry of troops and squadrons. A squadron consists of two troops, and a regiment, of two, three, or more squadrons.

Fortification. The art of defending and attacking fortresses and military positions, and of intrenching camps and outposts. It is either *natural* or *artificial*, *regular* or *irregular*. *Natural fortification* is the strength and security which nature has afforded to places (such as mountains, steep rocks, marshes, etc.,) by the advantages of situation and the difficulties of approach. *Artificial fortification* is contrived and erected to increase the advantages of a natural situation, and to remedy its defects. Fortification is *regular* when erected according to the rules of art, on the construction made from a figure or polygon which is regular, or has all its sides and angles equal; *irregular*, when the sides and angles are not uniform, equidistant, or equal, on account of the irregularity of the ground, rivers, hills, valleys, etc.

Fortifications (Subterraneous). Consist of the different galleries and branches which lead to mines, to the chambers belonging to them, and which are requisite when it is necessary to explode them for the purposes of attack or defense.

Fortifications (Field). Consist in the art of fortifying, constructing, attacking, and defending all kinds of temporary field-works during a campaign; as redoubts, field-forts, star-forts, triangular and square forts, heads of bridges, and various kinds of lines, etc.

Fort-Major. The commandant of a fort, in the absence of the regular commanding officer.

Fosse. A ditch, either with or without water in it.

Fougass. A small mine in the front of the weakest parts of a fortification; generally under the glacis or dry ditches.

Fraise. A row of stakes or palisades placed in an inclined position, on the edge of a ditch, or the outward slope of an earthern rampart. *To fraise a battalion*, is to line or cover it on every part with pikes or bayonets, to enable it to withstand the shock of a body of cavalry.

Funeral (Military). For the ceremonials of military funerals, see "Army Regulations," pages 34–35–36.

Furlough. Leave of absence granted to non-commissioned officers and soldiers.

Fuse. The tube of wood fixed to a loaded shell.

Fusil. A light musket or firelock.

Fusileer. A soldier armed with a fusil.

Gabion. A basket of a cylindrical form, filled with earth, either to carry on the approaches under cover during a siege, or in field-work. Parapets are often constructed of gabions. To construct gabions, some staves of the length of three or four feet are stuck into the ground, in the form of a circle, wattled together with osier twigs.

Gabionade. A retrenchment hastily thrown up. A parapet constructed of gabions is termed *a parapet en gabionade.*

Gallery. An underground passage leading to the mines. The term also is used for a communication between the interior and the exterior works of a fortified place.

Gantlet or Gauntlet. An iron glove. *To throw down the gauntlet*, in military acceptation, is to challenge; *to take up the gauntlet* is to accept the challenge.

General Officer. An officer above the rank of colonel.

Generalissimo. The chief officer in command in the field.

Glacis. The slope of the parapet of the covered way.

Grand Division. A body of troops composed of two companies. A regiment or battalion being told off in divisions of two companies, each is said *to be told off in grand divisions. Grand division firing* is when a battalion fires by two companies at the same time, and is commanded by only one officer.

Grappling Irons. Irons thrown at an object for the purpose of dragging it nearer.

Grenade. A hollow ball or shell of iron or other metal, about two and a half inches diameter, which being filled with fine powder, is ignited by means of a small fuse. It derives its name from having been formerly thrown by the grenadiers of regiments.

Guard. A body of men to protect an army or a place from being surprised. *The Advanced* or *Van Guard* is a party of cavalry or infantry, or of both arms, which marches before the main body, for the purpose of apprising it of any approaching danger. *A Rear Guard* is that part of an army or body of men which brings up the rear on a march, for the purpose of preventing the enemy from gaining ground on the flanks of the main body. The term is also applied to a corporal placed in the rear of a regiment, to keep good order in that part of the camp. *Main Guard,* that from which all other guards are detached. *Port Guard,* a guard detached from the main guard. *A Grand Guard* is a guard composed of three or four squadrons of cavalry, commanded by a field-officer, and posted about a mile from the camp for its better security. *An Advanced* or *Quarter Guard,* is a guard or detachment intrusted with the guard of a post. *Quarter Guard,* is a small guard posted in front of each battalion in camp. *Picket Guard,* see *Picket. Guards* are either *ordinary* or *extraordinary. Ordinary,* when mounted in camp or garrison towns; *extraordinary,* when detached to cover foragers, escorts, etc. *To Relieve Guard,* is to put fresh men or sentries on guard. *To turn out the Guard,* is to form the guard for the purpose of receiving a general or commanding officer; also, on the approach of an armed party, or on the beat of the drum, sound of trumpet, or any alarm. For the various details of guards and their service in the army of the United States, see pages 37–38–39–40–41.

Guard-Mounting. Is the hour at which a guard is mounted.

Guerrilla (Spanish for a little war). A partisan who is not enrolled and paid by the party for whom he serves. It is now, however, used to imply any irregular warfare.

Guides. Men who give information respecting the country, and the roads intersecting it.

Guidon. A cavalry standard or banner. Not used in our army parlance.

Gun (*Morning and Evening*). The gun fired every morning at sunrise, and every evening at sunset, to give notice to the drums and trumpets of troops in garrison, to beat and sound the reveille, and the retreat.

Gunnery. The science of artillery, or the art of managing cannon and military projectiles.

Gun-Shot. The reach or point-blank range of a gun.

Gymnastics. The art or method of exercising the body so as to render it supple, and capable of much fatigue. Much used in the Zouave drill, recently introduced.

Halberd or *Halbert.* A kind of spear formerly carried by sergeants of infantry and artillery. *Old Halbert,* a term once used in the army to designate a soldier who had risen to the rank of a commissioned officer. Not now in use.

Half-Pay. Allowance made to absent or retired officers.

Halting Days. Days allowed for repose when troops are on a march, and there is no necessity for exertion or dispatch.

Hang Fire (*To*). Fire-arms and trains of powder are said to hang fire, when a pause takes place between the ignition of the gunpowder and the application of the fire to it.

Herisson. A hedge or chevaux-de-friese, made of one stout beam, fenced with iron spikes, and fixed on a pivot, so that it revolves on being touched.

Hollow Square. The form in which a body of infantry is drawn up to resist a cavalry charge; with the colors, drums, baggage, etc., in the center.

Home Service. Military duty by citizens of towns, etc. *Regular Service,* is the performance of service in the army of the United States Government. *Foreign Service,* is service on a foreign station, beyond the limits of this country and its jurisdiction.

Honors of War. Terms granted to a capitulating enemy on evacuating a fortress.

Horn-Work. A kind of crown-work in advance of a fortress.

Hors-de-Combat (Put or Placed). Is to be killed, wounded, or disabled so as not to be capable of defense or attack.

Howitzer. A piece of ordnance for discharging shells at low angles, and shot in ricochet.

Hurdles. Oblong constructions of osier and willow twigs interwoven close together upon stakes for rendering batteries firm, or to consolidate a passage over muddy ditches, or to cover traverses and lodgments for the defense of workmen in trenches.

Impress Money. Money paid to men who have been compelled to serve. Not often used in our service.

Infantry. Foot-soldiers. The term having been applied to a body of men raised by an Infante of Spain, for the purpose of rescuing his father from the Moors; as a memorial of the deed the term was applied to foot-soldiers in general.

Inquiry (Board of). The meeting of a certain number of officers, for the purpose of ascertaining facts which may become matter of investigation by a court-martial.

Insconced. When a part of an army has fortified itself with a sconce, or small work, in order to defend a pass, etc., it is said to be *insconced.*

Intrenchment. A work which fortifies a post against attack. The term usually denotes a ditch or trench with a parapet. Intrenchments are sometimes made of fascines, with earth thrown over them, or of gabions, hogsheads, or bags filled with earth, as a protection from the enemy's fire.

Invalided (To be). Is to be discharged from the service in consequence of wounds, ill-health, or long service.

Invest. Investment. The investment of a fortress is the seizure of all the avenues leading to it, preparatory to its blockade or siege.

Judge Advocate. The public prosecutor of officers and soldiers tried by court-martial for breach of the articles of war or the general regulations.

Kit. The complement of a soldier's regimental necessaries.

Knights of the Round Table. A fraternity of twenty-four knights instituted by King Arthur. In order to prevent among them controversies about precedence, the King caused a round table to be made for them when assembled ; from which they were denominated *Knights of the Round Table.*

Lance Sergeant. A corporal who acts as sergeant, but receives only the pay of corporal.

Land Transport Corps. A body of men employed in conveyance of the wounded.

Laws of Arms. Certain acknowledged rules, regulations and precepts, which relate to war, and are observed by all civilized nations. The laws of arms also prescribe the method of proclaiming war and commencing hostilities.

Law (Military). A prompt and decisive rule of action by which justice is dispensed to the public or to individuals, without passing through the tedious channels of legal investigation.

Leading Column. The first column which advances from the right, left, or center of an army or battalion. *The Leading File,* the first two men of a battalion or company, that marches from right, left, or center, by files.

Levy (To). Has three distinct military acceptations ; to levy or raise an army, to levy or make war, and to levy contributions.

Lie under Arms (To). To be in a state of preparation for action.

Light Bobs. A familiar term for light infantry.

Light Infantry. A company of the active, strong men of a battalion. A regiment employed as light infantry is divided into skirmishers, supports and reserve. The supports are in the rear of the skirmishers. The reserve is the point on which both the supports and skirmishers may rally.

Limber. The fore-part of a traveling gun-carriage, to which it is fastened by means of a pin-tail or an iron pin. The hooking or unhooking the gun or howitzer-carriages from the limbers is called, in the artillery service, for retreat or advancing, *limbering up,* that is making every thing ready in the gun-carriage ; and for action, *unlimbering.*

Line of Battle. The arrangement or disposition of an army for battle.

Line of Communication. In military strategy, that line which corresponds with the line of operation, and proceeds from the base point. The term also denotes that space of ground which unites the citadel to the town.

Line of Direction. In gunnery, a line formerly marked upon guns, to direct the eye in pointing the gun.

Line of Fire. The space between contending armies, or any space from which objects may be hit by cannon or musketry.

Line of March. The regular and tactical succession of the component parts of an army in motion. The term also signifies the distance of ground over which armed bodies of men move in succession toward a given object.

Line of Operation. The line which corresponds with the line of communication, and proceeds from the base point; or the forward movements of an army, for the purpose of attacking the enemy, penetrating his dominions, etc.

Line (To) Men. Officers and non-commissioned officers are said *to line the men* belonging to their several battalions, divisions, or companies, when they arrive at their dressing points, and receive the word *dress* from the commanding officer.

Line (To Form the). To arrange troops in order of battle, or battle array.

Line (To Break the). To attack an opposing front, so as to throw it into confusion. The term also signifies to change the direction from that of a straight line, for the purpose of obtaining a cross fire.

Lines of Approach. See *Trench.*

Lines of Communication. The trenches which unite one work to another, so as to insure communication between two approaches at a siege, or between two posts or forts.

Lines of Intrenchment. Lines which are drawn in front of a camp, or a place indifferently fortified, to secure it from assault or surprise.

Lines of March. Bodies of armed men marching on given points for the purpose of arriving at any straight alignments on which they are to form.

Lines of Support. Lines of attack which are formed to support one another.

Lines. A series of field-works, either continuous or at intervals, contrived so as reciprocally to flank one another. When continuous they are termed *Continual Lines;* when at intervals *Interrupted Lines,* or *Lines with Intervals.*

Links (*Connecting*). The men sent out from a support, to keep up its connection with the skirmishers.

Lodgment. A retrenchment made for shelter in a captured post or outwork, for the purpose of maintaining the position. The term also signifies the possession of an enemy's work.

Loop-holes. Openings in the walls of a castle or fort, through which the garrison may fire. In general, they are nine inches long, six or seven wide within, and two or three without, so that there may be a direct fire from them in front, or an oblique fire to right or left, according to circumstances.

Lying. In military parlance, signifies to be stationed or quartered in a given place.

Magazine. A place in which military stores, arms, ammunition, provisions, etc., are deposited.

Main-Body. The body of troops which march between the advanced and rear guards. In a camp, it is that part of the army which is encamped between the right and left wings.

Main Guard or *Grand Guard.* A body of cavalry posted in front of a camp for the security of the army. In garrison, it is a guard mounted generally by a subaltern officer and twenty-four men.

Malingerer. A soldier who feigns illness to avoid his duty.

Mammelon. A round hillock of easy ascent, rising upon the surface of the ground. The word in French literally signifies a nipple.

Maneuvers. Military evolutions. *To maneuver troops,* is to habituate them to a variety of evolutions, accustom them to different movements, and to render them familiar with the principles of offensive and defensive operations. The term also signifies the management of an armed force, so as to derive sudden and unexpected advantages before an enemy.

Manual Exercise. A regulated method of rendering troops familiar with the musket, and of adapting their persons to military movements under arms. *Platoon Exercise* is the method of drilling soldiers in small numbers or subdivisions. *Sword-bayonet Exercise* is that in which riflemen are taught to use their swords when fixed to the rifle.

Mantlets. Wooden fences, on rolling wheels, used during a siege to protect the sappers from the enemy's fire.

March (Dead). The strains of music played during the procession of a military funeral.

March (Rogue's). The beats of a drum when a criminal offender is expelled or drummed out of a regiment.

Marching. Is either in slow time, quick time, or double-quick time. In *slow* or *quick time,* the length of the step or pace is thirty inches, except in stepping out, when it is increased to thirty-three inches, and in stepping short, when it is reduced to ten inches. In *double-quick time,* the step or pace is thirty-six inches. In *slow time,* seventy-five steps or paces are taken in a minute; in *quick time,* one hundred and eight, and in *double-quick time,* one hundred and fifty. The *side* or *closing step,* which is taken when it is necessary to move a small distance to either flank, is ten inches, and is always taken in quick time; but when taken to clear or cover another soldier, it is twenty-one inches. In *stepping back,* the step or pace is thirty inches.

Marching by Files. To march with the narrowest front, except that of rank entire or Indian file, of which bodies of troops are susceptible.

Marching, or *Billet Money.* Money paid to officers and soldiers, for covering their expenses incurred when marching for the purpose of changing quarters.

Marching Regiments. A term given to those corps who have not any permanent quarters. Latterly, they have been denominated *regiments of the line* or *line regiments.*

Marines or *Marine Forces.* Troops raised for the naval service, and trained to fight either in a naval engagement, or in an action on shore.

Mark (To) Time. Is alternately to throw out each foot, bringing it back square with the other, without gaining ground, so that the cadenced step may be preserved until the obstacle is removed which required the necessity of marking time. *Changing the feet in marching,* by quickly bringing up the ball of the rear foot to the heel of the advanced one, and instantly making another step forward, is to recover the cadenced step which has been lost.

Martello Towers. Round towers of a conical form, rather broader at the base than the summit, and about forty feet in height, constructed as coast defenses. Not now used in our military arrangements.

Martial Law. The law of war.

Martinet. A strict disciplinarian, who gives officers and soldiers unnecessary trouble. The term is supposed to have had its origin from an adjutant of that name, who was in high repute, as a drill-officer, in the reign of Louis XIV.

Masked. Concealed.

Matériél. The appurtenants of an army, such as horses, cannon, ammunition, stores, provisions, etc.

Melée. A confused hand-to-hand fight.

Mess. A kind of ordinary or table d'hote, at which the officers of a regiment dine. For the mess and regulations of the *Soldiers' Mess,* see pages 30–31.

Military Fever. Humorously called the Scarlet Fever; an overweening fondness for the outward appendages of the soldier.

Military Messengers. Confidential men who are sent on messages, or with letters to and from head-quarters, etc.

Military Regulations. The rules and regulations by which the discipline, formations, field-exercise, and movements of the army are directed to be observed, according to a uniform system.

Militia. Citizen soldiers. Each State has its separate military organization, by which all citizens liable to bear arms are enrolled, and mustered into divisions, brigades, regiments, companies, etc. In most States, even the commissioned general officers are elective, each division electing its commanding officers, who are commissioned by the Governor on their certificates of election. The Governor nominates the adjutant-general, quartermaster-general, inspectors, etc., as well as his own staff. The militia, in time of peace, are only required to attend the various annual or semi-annual musters ordered by State laws. In event of war, the general Government issues its proclamation announcing war and its causes, when the Secretary of War makes his requisition on the Governors of the States for the needed troops. The Governors then accept volunteers for their quota, and send them forward at command of the War Department. If volunteers do not offer in sufficient numbers, then the requisite force is obtained by *drafting.* The militia are mustered, and the white and black bean, drawn from a box, indicates the men. These either go, or find substitutes—there is no other course. By this admirable system our Government is saved the expense of a large standing army, yet has, upon short call, several millions of troops. The facility of the organization has been made manifest in our recent internecine troubles. An army of very effective men, 250,000 strong, was ready for the field in six weeks' time.

Minié Rifle. A rifle invented by Captain Minie, of France, carrying a conical ball, hollowed at its base. The powder exploding, *expands* the base of the ball closely into the grooves of the rifle-barrel. Additional force is thus gained for the ball.

Mining. The making of subterraneous passages under the wall or rampart of a fortification, for the purpose of blowing it up with gunpowder. *Counter-mining*, the making of galleries and mines by the besieged, to counteract the mines of the besiegers.

Missing. The expression used in military returns, especially in field-reports, after an engagement, to account for the general loss of men.

Mitraille. Small pieces of old iron, as heads of nails, etc., with which pieces of ordnance are loaded, commonly called grape-shot.

Mobilize (To). To embody or incorporate. Used in the French service vocabulary.

Mortars. Short brass or iron cannon, of a large bore, for throwing shells.

Mount (To) Guard. To go on duty.

Mount (To) Cannon. To put a piece of ordnance on its frame, for its more easy carriage, and the management of it in firing. *To dismount cannon*, is to remove it from its serviceable position.

Mount (To) a Gun. Is either to put the gun into its carriage, ōr when in the carriage, to elevate the mouth, or raise it higher.

Movements. In military parlance, signify the different evolutions, marches, counter-marches, and maneuvers made in tactics.

Musketry Range. Effective musketry range of fire is when delivered against infantry from 200 to 250 yards, and against cavalry from 30 to 60 yards. The reasons that the range of a cannon or a musket-ball is limited when fired from rifled artillery and musketry, or from the smooth-bored cannon and musket, are, that its momentum gradually diminishes in its range or flight, and is subject to the friction of the air in its passage through it. A ball fired in vacuo would have thirty-four times the range which it has when fired in air. Another reason that musketry fire does so little execution when delivered by troops aligned, is, that owing to the curvature of the earth, at the distance of 800 yards, a man of ordinary stature, presents a mark of only *one-tenth of an inch in altitude or height*, and at 1,000 feet, but little more than a twelfth. Colonel Schliminback, of the Prussian service, from a number of calculations extending over a series of battles during the wars which sprung out of the French Revolution, ascertained that a man's weight in lead, and ten times his weight in iron, are consumed before he is put· *hors de combat!* At the battle of Vittoria, nearly four millions and a quarter of ball cartridges and 6,570 round shot and shell were fired by artillery, but the killed and wounded of the French army, consisting of 90,000 men, did not amount to 8,000. The same was true in the Crimean struggle. Notwithstanding the terribly destructive nature of the contest, it is proven that two hundred and eighty shots were expended for every man killed! This makes no account of the vast amount of shot and shell used. If even one in ten shots proved fatal, an army would soon be annihilated.

Muster-Roll. A nominal return of the officers and men in every regiment, troop, or company in the service, forwarded monthly to the War Department.

Mutiny Act. A statute specifying military offenses, and by virtue of which the English army is continued on a peace or a war establishment.

Naval Camp. A fortification consisting of a ditch and a parapet on the land side, or a wall built in the form of a semi-circle, and extending from one point of the sea to the other.

Necessaries (*Regimental*). The boots, shirts, stockings, *et cetera.*, issued to soldiers.

Non-commissioned Officers. Are those officers elected by the men, or those not served with commissions, either by the general Government or by the State authorities. See *Officers.* In infantry regiments are, the sergeant-major, the quartermaster-sergeant, the sergeants, corporals, and drum and fife majors.

Non-effective. The negative of *effective.*

Oblique (*To*). To move forward to the right or left, according to the word of command.

Oblique Deployment. Is when the component parts of a column extending into line, deviate to the right or left, for the purpose of taking up an oblique position : in which operation its movements are termed *oblique deployments.*

Oblique Fire or *Defense.* A fire under too great an angle.

Oblique Step. A step or movement in marching, taken gradually to the right or left, at an angle of about twenty-five degrees.

Obstacles. In a military sense, are narrow passes, or any impediments which present themselves when a battalion or other body of men is marching to front or rear.

Officers. Are commissioned or non-commissioned. *Commissioned Officers* are either general, field, staff, or subaltern. *Staff Officers* are the quartermaster-general, the adjutant-general, brigade officers, and aides-de-camp, etc.

Off-Reckonings. The account of money issued by Government to colonels of regiments, for the clothing of the men.

Opening of Trenches. The first breaking of ground by the besiegers, for the purpose of carrying on their approaches toward the place.

Order of Battle. The arrangement or disposition of the various component parts of an army for battle.

Orderly. A non-commissioned officer or private who attends an officer for the performance of orderly duty.

Orderly Book. A book into which the sergeants of companies transcribe the general and regimental orders, for the specific information of the officers and men.

Orderly Officer. The officer of the day.

Orderly Room. A room in barracks used as a regimental office.

Orders. In a military acceptation, are the lawful commands of *superior* officers relative to military affairs, and are: *General Orders,* which are those issued by the commander-in-chief, for the government of the army at large, or for any specific purpose : Abbreviated *G. O. Garrison Orders,* those issued by the governor of a garrison : Abbreviated *Gar. O. District Orders,* those issued by a general commanding a district : Abbreviated *D. O. Brigade Orders,* those issued by a general commanding troops brigaded : Abbreviated *B. O. Regimental Orders,* those issued by the commanding officer of a regiment, arising out of general or garrison orders: Abbreviated *R. O. Standing Orders,* general rules and instructions which are to be invariably followed, and are not subject to the temporary intervention of rank. Of this description are those orders which the colonel of a regiment may judge fit to have inserted in the orderly books, and which can not be altered by the next in command, without the colonel's concurrence: Abbreviated *S. O. Station Orders*, orders issued by the commanding officer of a particular station or post for its interior government: Abbreviated *Sn. O. Pass Orders*, written directions to sentries, etc., belonging to outposts, etc., to allow the bearer to go through the camp or garrison: Abbreviated *P. O. Beating Order*, an authority given to an individual, empowering him to raise men by beat of drum, etc., for a particular regiment, or for general service.

Ordnance. Heavy artillery, as cannon, howitzers, mortars, etc.

Outfit. The necessaries, uniform, etc., which an officer provides when appointed to a commission.

Outpost. A body of men posted beyond the grand guard, or the limits of a camp.

Outworks. The works constructed beyond the *enceinte* or body of a place, as ravelins, half-moons, tenailles, horn and crown-works, lunettes, etc.

Pace. The military step. The word also signifies the relative distance in the formation of a battalion at close or open order.

Palisades. Wooden stakes, about nine feet long, and six or seven inches square, having one end sharpened in a pyramidical form to the extent of a foot. They are planted three feet deep in the ground. When placed in an inclined position, they are termed *fraises.*

Parallels. The trenches which connect the approaches and batteries carried on before a besieged fortress.

Parallel Lines. Lines drawn in the same direction, preserving equal distances from each other.

Parapet. A screen of a fortified post to protect troops from the enemy's fire.

Park of Artillery. A spot in an encampment in which the artillery is placed. The term also signifies the whole train of artillery *matériél* belonging to an army in the field.

Parley. A conference with an enemy by means of a flag of truce.

To Beat a Parley. Is to give a signal for holding a conference by beat of drum, or sound of trumpet.

Parole. The promise or word of honor given by a prisoner of war, when he has leave to go at large, of returning at an appointed time, or not to take up arms, if not exchanged. A paroled person, if taken with arms in his hands, is shot for violation of his parole.

Partisan. One dextrous in commanding a party for obtaining intelligence, surprising the enemy's convoys, etc. The term is used to denominate an officer who has the command of a partisan corps or party. It is also used, in American parlance, to signify a guerrilla leader—one who serves with no particular division or regiment.

Party. A small number or detachment of men, either cavalry or infantry. *Recruiting Party*, a certain number of men, under an officer or a non-commissioned officer, detached from their battalions for the purpose of raising recruits. *Working Parties*, small detachments of men, under the command and superintendence of officers, who are employed on fatigue duties.

Passages. Openings cut in the passages of the covered way, to afford communication to all its parts.

Patrol. A small party under the command of a subaltern or a non-commissioned officer, detached from the main or quarter guard, to patrol, for the purpose of maintaining order and regularity in a camp or garrison.

Pause. The stop or intermission between the first and last words of a command.

Pay-Sergeant. The non-commissioned officer who pays the men of each company their pay.

Peace-Establishment. The reduced number of regulars when the country is in a state of peace.

Permanent Rank. In the army, which does not cease with a particular service or locality of service ; a term in contradistinction to *local* or *temporary rank*, which ceases on the performance of the duty for which it was granted. Thus, officers in the regular service are of permanent rank — those in the volunteer or militia service are of temporary rank.

Petard. A pot charged with gunpowder, fixed against the gate of a fortress, for the purpose of blowing it open. In recent practice, leather bags containing powder have superseded the use of the petard.

Pickets. Sharp stakes for securing the fascines of batteries or fastening the tent ropes of camps. *Picket Ropes* are ropes twisted at given intervals round the several picket stakes, to confine the horses within a proper space of ground. *Picket Poles* are round pieces of wood, shod with iron, and driven firmly into the earth, to fasten the cavalry horses by, when at picket.

Picket. A small detachment of cavalry or infantry, which is either *in-lying* or *out-lying*. *An In-lying Picket* is within the lines of intrenchment of a camp, or within the walls of a garrison town, ready to turn out on alarm. *An Out-lying Picket*, is that which does duty without the limits of a camp or garrisoned town; being in the first-mentioned position, posted on the front and flanks of the army, to guard against surprise, or to oppose reconnoitering parties. They are also called *In-Line* and *Out-Line* pickets.

Pike. A shaft of wood, from ten to fourteen feet in length, pointed with a flat steel blade, about six inches in length. Men armed with pike, cutlass, and revolver are very formidable on charges, or in close conflict. In event of the negroes of the slave States rising in insurrection, almost their only reliance would be the pike, which they themselves would manufacture. The pike is much used in South American warfare. As a *lance*, it is used both by horse and foot soldiery. A regiment of pikemen is a very desirable organization for every grand army.

Piling Arms. Locking muskets together by means of that part of the ramrods near the muzzles of the pieces; and *Unpiling Arms*, is the unlocking or detaching them from one another.

Pioneers. Soldiers selected from every regiment for mending the ways, removing obstacles, working on intrenchments and fortifications, and for making mines and approaches.

Pivot. The officer or soldier stationed at the flank of a company, on whom the different wheelings are made.

Place of Arms. When an army takes the field, every stronghold or fortress which supports its operations by affording a safe retreat to its depots, heavy artillery, magazines, hospitals, etc., is called a *place of arms.* In offensive fortifications, those lines are called *places of arms* or *parallels*, which unite the different means of attack, secure the regular approaches, etc., and contain bodies of troops who either do duty in the trenches, protect the workmen, or are destined to make an impression on the enemy's outworks.

Platoon. A term implying a subdivision of troops, either less or more than a company, who act in concert of fire.

Point of Alignment. The point on which troops form and dress by.

Point Blank Range. Is when a cannon or musket is leveled horizontally, so that the muzzle neither mounts nor sinks, but that the surface lines of the piece and the object are in the same plane.

Police Guard. A regimental guard, detailed every day, commanded by a lieutenant that furnishes ten sentinels for special duty.

Pontoon. A kind of vessel hull formed of wood, and covered with copper, for the purpose of forming temporary bridges to cross rivers.

Pontooning. The art of constructing a temporary bridge by means of pontoons.

Pontoon Train. The whole equipment requisite for pontooning.

Portfire. Paper-cases filled with saltpeter, sulphur, and mealed powder, to serve as a slow match for artillery.

Post. A spot of ground, fortified or not, where a body of men can be in a condition to resist an enemy. *An Advanced Post* is a spot of ground seized by a party to secure its front and the post behind it. *Post of Honor*, an honorable position. The advanced guard is the post of honor, and the right of the two lines is entitled to the same distinction.

Prestige. Illusion, charm, moral force.

Profile. The drawing of a section of a parapet, or other work, sideways.

Projectiles. Shot or shell discharged from artillery.

Provost-Martial. An officer appointed to preserve good order and discipline, apprehend offenders, and superintend their punishment.

Punishment. In military usage means any infliction of sentence for dereliction of duty or transgression, and consists of arrests, confinements, deprival of arms, cashiering, drumming out of camp, etc. In cases of treason, desertion and spying, the penalty is death by shooting or hanging. See "Army Regulations," Article 27.

Pyrotechny. In a military acceptation, is the manufacture of bombs, grenades, rockets, fire-lights, etc.

Quarters. Military stations, as head-quarters, home-quarters, regimental-quarters, etc.

Raid or *Razzia.* A plundering or marauding incursion.

Raise (*To*) *a Siege.* To abandon the siege of a fortress.

Rally (*To*). To re-form troops disordered or dispersed. *A Rallying-Square,* is a square formed around an officer by skirmishers surprised by cavalry.

Rampart. The exterior elevation of a fortified place upon which guns are placed in position.

Random-Shot. Is when a piece is elevated an angle of forty-five degrees upon a level plane.

Ranging. Disposing troops in proper order for battle, maneuver, march, etc.

Rank and File. The horizontal and vertical lines of troops when drawn up.

Rappeler. A particular beat of the drum to recall soldiers to the defense of their colors.

Rations. A certain allowance either for officers or men, given in bread, meat, or forage, when troops are on service.

Ravelin or *Demilune.* A work constructed on the counterscarp before the curtain of a fortress.

Ravine. In field fortification, a deep hollow.

Raw. In military acceptation, unseasoned, wanting knowledge in military tactics, etc. *Raw Troops,* inexperienced soldiers, who have been but little accustomed to the use of arms.

Razed. Works and fortifications when demolished, are said to be razed.

Recoil. Or, as it is properly termed, *the Kick,* is the rebound or backward motion which a cannon or gun takes from the explosion of an overcharge of powder.

Reconnoissance. The act of reconnoitering an enemy's position.

Reconnoiter (*To*). To view and examine a position. *Balloon reconnoitering* is by means of balloons.

Recruits. Men raised on the first formation of a corps, or to supply the places of those who have been disabled or killed.

Recruit Horses. The horses for completing regiments of cavalry.

Redans. In field fortification, are indented works, lines, or faces, forming sallying and re-entering angles, flanking one another, generally constructed on the side of a river running through a garrison town.

Redoubts. Works about musket-shot from a fortress, surrounded by a ditch. *Field Redoubts* are temporary defenses or fortifications, thrown up during a war of posts, or on account of sudden emergency.

Re-form (*To*). In a military acceptation, is, after some maneuver or evolution, to bring a line to its natural order, by aligning it on a given point.

Refuse (*To*). To throw back, or keep out of that regular alignment which is formed when troops are on the point of engaging an enemy.

Relief. A fresh detachment of troops who replace those on duty.

Relieve (*To*) *the Guard.* Is to put fresh men on guard. *To relieve the trenches,* is to relieve the guard of the trenches, by appointing those for that duty who have not been there already, or whose turn is next. To relieve the sentries is to put fresh men from the guard on that duty.

Rendezvous. The place appointed for the assembly of a body of troops in case of alarm.

Reserve. A select body of troops retained in the rear of an army, to support the attacking force, or to rally it in case of disaster.

Respited (*To be*) *on the Muster Roll.* Is to be suspended from pay, etc., during which period all advantages of promotion, pay, etc., are stopped. Not much used in our army parlance.

Retreat. The retrograde movement of an army or body of men. To be in full retreat is to retire expeditiously before the enemy.

Retrenchment. A work raised to cover a post, and fortify it against an enemy; such as fascines loaded with earth, gabions, barrels, etc., filled with earth, sand-bags, and generally all things that can cover the men, and stop the enemy; but it is more applicable to a ditch bordered with a parapet; and a post thus fortified, is called a *retrenched post,* or *strong post.*

Revêtement. A strong wall built on the outside of the rampart and parapet, to support the earth, and prevent its falling into the ditch.

Revolver. Fire-arms which produce a series of successive discharges from the chambers of the barrel of a single arm or stock.

Ricochet, Boundings or leaps of round shot. *Ricochet-Firing,* is firing at a slight elevation, in a direction enfilading the face of a work; so that when the shot falls over the parapet, it makes several bounds along the rampart, with destructive effect on the guns and gunners.

Rideau. A rising ground, or eminence, commanding a plain.

Rifle. A firelock of which the bore is furrowed or grooved in a spiral or screw-like form. The rifles of the highest repute are the Enfield, Minié, Sharpe's, Whitworth's, Colt's repeating rifle, etc. See article on "Rifles," in the body of this work.

Riflemen. Expert marksmen, armed with rifles.

Rifle Pits. Pits in which riflemen ensconce themselves, to pick off the gunners at the embrasures of a fortress.

Rocket. A firework, used either as a signal or a projectile.

Roll. A uniform beat of the drum.

Roll-Call. The calling over the names of the men. .

Roster. A plan or table by which the duty of officers, battalions and squadrons is regulated. See "Army Regulations," pages 73-74.

Round. A general discharge of cannon or musketry.

Round of Ammunition. The number of ball-cartridges with which a soldier is supplied.

Round-Robin. A compact of honor which officers enter into, (when they have cause of complaint against their superior officer), to state their grievances, and to endeavor to obtain redress, without subjecting one individual more than another to the odium of being a leader, or chief mover. The term is a corruption of *ruban rond*, which signifies a round ribbond.

Rounds (Visiting). The visitation and personal inspection of guards and sentries on duty.

Rounds (Grand). The rounds which are gone by general officers, governors, commandants, or field-officers.

Route. The order for the march of a regiment or detachment, specifying its various stages.

Ruffle. A vibrating sound made by drummers on the drum, not so loud as the roll.

Running Fire. A rapid succession of fire.

Safeguard. A protection granted for the preservation of an enemy's lands or persons from insult or being plundered.

Salamanders and Serpents. Brittle earthern vessels, filled with serpents, which were thrown among a storming party on the point of ascending a breach, for the purpose of annoying them.

Sally or *Sortie.* A sudden attack made by the besieged against the troops or works of the besiegers.

Sally-Ports. Openings in the glacis of a fortress, for the purpose of egress and regress of troops engaged in a sally or sortie.

Salute. A discharge of artillery or musketry, or of both, in honor of persons or events. The term also signifies the ceremony of presenting arms.

Sap. A trench or approach sunk underground to protect the workmen from the fire of the garrison. A flying sap is that in which the working parties place the gabions themselves, and instantly fill them themselves.

Sapping. The method of carrying on the approaches, by excavating trenches so as to protect the workmen from the fire of the garrison.

Sappers. Soldiers belonging to the corps of artificers and engineers who work at the saps. Sapping party, the men who form a brigade or party of sappers.

Saucisse, Saucisson. In mining, a long pipe, or bag, made of cloth, well pitched, or sometimes of leather of one and a half inch in diameter, filled with powder, laid from the chamber of a mine to the entrance of the gallery. The term also signifies a kind of fascine, longer than those usually made, for the purpose of raising batteries, or repairing breaches. They are also used in making epaulements, stopping passages, making traverses over a wet ditch, etc.

Scaling Ladder. A ladder for scaling or mounting walls or ramparts.

Scarp (To). To render a slope accessible by cutting it down.

Sconce. A redoubt or small fort.

Scour (To). A term to express the act of firing a quick and heavy discharge of ordnance or musketry for the purpose of dislodging an enemy; as to scour the rampart or the covered way. The expression also signifies to clear, or to drive away, as to scour the streets, to scour the trenches, etc. *To scour a line,* is to so flank it so as to be able to see along it, and that a musket ball discharged at one end, may range to the other.

Scouts. Men employed to gain intelligence of the forces and movements of an enemy.

Sentinel, Sentry. A soldier posted to watch the approach of an enemy, prevent surprise, or to stop and challenge those who approach his post.

Sergeant (Covering). A non-commissioned officer, who during the exercise of a battalion, regularly stands or moves behind each officer commanding or acting with a company.

Shells. Hollow iron balls, filled with powder, thrown from mortars or howitzers. Message Shells are shells in the inside of which a letter or other papers are put.

Shot. A denomination given to all kind of balls used for artillery and fire-arms, whether round, grape, chain, case, or canister.

Siege. The position which an army takes, on its encampment, before a fortified town, or place, for the purpose of reducing it. The first operation is investing the place, that is, taking possession of all the avenues, forming lines of circumvallation, opening the trenches, etc. In siege-operations, *the rear of an attack* is the place where the attack begins; *the front or head of the attack,* that part next to the place.

Sight. A small piece of brass or iron, which is fixed near the muzzle of a musket or pistol, to serve as a point of direction, and to assist the eye in leveling. Rifles have two sights, one at the breach, and the other at the muzzle; and some rifles have telescope sights.

Signal. A sign for conveying intelligence by balls, rockets, or flags. Signals are also given by the short and long rolls of the drum during the exercise of a battalion. They are frequently given for commencing a battle, either with drums and trumpets, sky-rockets, the discharge of cannon, etc. The term is also used for an order for marching, etc. Secret signals are in frequent use in our army, both for conveying orders and intelligence. A new "code" has recently been adopted by General Scott. "Codes" change as often as necessary to preserve their secrecy.

Signal-Staff. A flag planted upon the spot where the general, or commanding officer, takes his station.

Size-Roll. A list containing the names of the men composing a troop or company, with the height or stature of each specified.

Soldier's Thigh. A figurative expression for an empty purse or a pair of trowsers which set close and look smooth, because they have no pockets, or nothing in them.

Sobriquets (*Regimental* and *Divisional*). Are cognomens obtained for some special conduct or circumstance. The habit of naming regiments has become very general. We have, for instance, "The Advance Guard," "The Garibaldi Guard," "The Union Guard," "The Scott Guard," "The President's Guard," "The Invincibles," etc.

Spike (*To*) *Cannon*. To drive a large nail or iron spike into the vent.

Squad. A small number of men, cavalry, or infantry, who are collected together for the purposes of drill, etc. *To squad*, is to divide a troop or company into parts, for the purpose of drilling the men separately, or in small bodies. *The awkward squad*, consists not only of recruits at drill, but of soldiers who are ordered to exercise with them, in consequence of some irregularity while under arms.

Squad-Roll. A list containing the names of each squad.

Squadron. A body of cavalry, composed of two troops.

Staff. In military acceptation, is either general, personal, regimental, garrison, or district. *A General Staff* consists of a quartermaster-general, an adjutant-general, majors of brigade, aides-de-camp, etc. *A Personal Staff* consists of those officers who are constantly about the person of a governor or a general, as his military secretary, aides-de-camp, etc. *A Regimental Staff*, are the adjutant, paymaster and surgeon.

Station (*Military*). A place for the rendezvous of troops. The term is also used to designate the spot for offensive and defensive measures.

Step. The pace of the soldier while marching in slow, quick, or double-quick time. *Stepping Out*, is lengthening the pace to thirty-three inches, by leaning a little forward, without altering the cadence. *Stepping Short*, is taking but ten-inch paces. In *Stepping Back*, the pace and cadence are the same as in the slow march. *The Diagonal Step*, is carrying the left foot forward nineteen inches in the diagonal line to the left. *The Balance Step*, popularly termed *the goose step*, is alternately throwing out the feet, without gaining ground.

Stockade. A work consisting of palisades.

Stock Purse. A saving made in a corps, and which is applied to regimental purposes.

Stores (Military). Provisions, forage, clothing, arms, ammunition, etc.

Stormers. The troops who immediately follow the forlorn hope in the assault of the breach made in the walls of a fortress.

Strategy, Strategetics. The science of military command, and of planning and directing military movements.

Subaltern. An officer under the rank of captain. The term, in familiar expression, is abbreviated *Sub.*

Subdivision. A company told off for parade or maneuver into two equal parts.

Supernumerary Officers, and *Non-Commissioned Officers.* Those placed in the rear for supplying the place of those who fall in action, and for preserving order and regularity in the rear ranks while the front rank is engaged or is advancing.

Swivel. A small piece of ordnance which turns on a pivot or swivel.

Tactics. The arrangement and formation of troops by means of maneuvers and evolutions.

Take (To) Ground to the Right or Left. To extend a line toward either of those directions.

Tambour. A work formed of palisades.

Tampions or *Tompions.* Wooden cylinders to put into the mouths of guns, howitzers and mortars in traveling, to prevent dust or wet entering them.

Target. A mark employed in the practice of ball-firing.

Tarred (To be). A cant expression in use among the regiments of guards, to signify the punishment which privates undergo among themselves, when they have been tried and sentenced by their comrades.

Tirailleurs, Voltigeurs. In the French service tirailleurs are skirmishers or marksmen, advanced in front to annoy the enemy, and draw off his attention; or they are posted in the rear to amuse and impede his advance in pursuit. Voltigeurs (springers, leapers), are employed for the same purpose. The distinctive employment of tirailleurs and voltigeurs is, that the first move irregularly and scattered; the second are formed and act in collective bodies.

Tire. Large guns, shot, shells, etc., placed in a regular form.

Tell (To) Off. To divide and practice a regiment or company in the several formations, preparatory to marching to the general parade for field-exercise. A regiment is told off into wings, grand divisions, divisions or companies, and subdivisions or sections. A company is told off into subdivisions and sections.

Terre-Plein. In field fortifications, is the plane of a level country around a work ; in permanent fortification, it signifies the broad surface of the rampart, which remains after constructing the parapet and bauquette.

Tête-de-Pont. A field-work or fortification, in front of a bridge, in the form of a redan, a system of crémailleres, a horn or crown-work, or portions of star or bastioned forts.

Touch (The). In a military sense, signifies the sensation felt by the soldier, when properly in line, at the thick part of the arm, immediately below the elbow, and which is communicated to him by his right or left-hand man, according to the point of direction in which the line is marching.

Tour of Duty. Duty by turn or succession.

Traverse. In siege operations, is a kind of retrenchment made in the dry ditch to defend the passage over it. *To Traverse a Gun or Mortar*, is to bring it about to right or left with handspikes, till it is pointed exact to the object.

Train. All the necessary apparatus and implements of war, as cannon, etc., required at a siege or in the field. *Train of Artillery*, the ordnance belonging to an army in the field. *Field Train*, a body of men, consisting chiefly of commissaries and conductors of stores, who belong to the artillery. *Train*, in mining, a line of gunpowder laid for the purpose of blowing up earth-works, etc.

Trenches. Passages or excavations made by besiegers, in order to approach more securely to the place attacked, on which account they are also termed *Lines of Approach*. *The tail* in rear of the trench is the place where it was begun; *its head* is where it ends. *Returns of a Trench* are elbows and turnings which form the lines of approach. *To open the Trenches*, is to break ground for the purpose of carrying on approaches to the place. *To mount the Trenches*, is to relieve the guard of the trenches. *To scour the Trenches*, is to make a vigorous sally on the guard of the trenches, force them to give way, and quit their ground, drive away the workmen, break down the parapet, fill up the trench, and nail or spike the cannon. *Counter-trenches*, are trenches made by the besieged against the besiegers. Trenches are also made to protect an encampment.

Troop. A company of cavalry. *Trooper,* a horse soldier.

Trouee. An opening, a gap.

Trous-de-Loup, or *Wolf-Holes.* In field fortifications, are round holes about six feet deep, and four feet in diameter, pointed at the bottom, with a stake planted in the middle. They are frequently dug round a redoubt, to obstruct the enemy's approach.

Tumbrils. Covered carts, which carry ammunition for cannon, tools for the pioneers, miners and artificers; and sometimes the military chest.

Truncheon. A staff of command.

Trunnions. The arms by which a gun is attached to its carriage.

Turn (To) out the Line. To exhibit in battle array, men for the purpose of parade, or to bring them into action.

Under Arms. Troops are under arms when assembled, armed, and accoutered, on parade.

Upshot (To). To extract a ball from a piece.

Van-Guard. That part of an army which marches in front.

Vedettes. Mounted sentries, stationed at the outposts of an army or an encampment.

Vent. The passage or opening in fire-arms, by which the fire is communicated to the powder composing the charge.

Visiting Officer. An officer whose duty it is to visit the guards, barracks, messes, hospital, etc., for the purpose of noticing whether the orders or regulations which have been issued respecting those matters are observed.

Volley. The simultaneous discharge of a number of fire-arms.

Volunteers. Those who enter the service of their own accord.

Wad. In gunnery, a substance made of hay or straw, and sometimes of tow rolled up tight, in the form of a ball.

Wadding. Hay or straw, or any other kind of forage, carried along with the guns to be made into wads.

War (Council of). An assembly of officers convened by a general to deliberate with him on enterprises, etc. The term is also used to designate an assembly of officers sitting in judgment on delinquent soldiers, deserters, cowardly officers, etc.

War-Cry. A cry formerly customary in the armies of most nations, when they were on the point of joining battle. Sometimes it consisted of tumultuous shouts, or horrid yells, uttered with an intent of striking terror into their enemies. At the battles of Crecy, Poictiers, and Agincourt, the war-cry of the English was "God and St. George," that of the French, "Montjoe and St. Denis." In our service, each regiment has its war-cry.

War Establishment. The number of effective men who compose the army in time of peace.

War (*Seat* or *Theater of*). The extent of country in which war prevails.

Watchword. The word given out in the orders of the day, in time of peace; but in time of war, every evening in the field, by the general who commands, and in garrison by the governor, or other officer commanding in chief, to prevent surprise, or the ingress and regress of a spy; it is generally termed *the parole* or *sign*, and to which is added the *countersign.*

Wheel (*To*). In a military acceptation, is to move forward or backward in a circular manner, round a given point.

Wheelings. Different motions made by cavalry and infantry, either to the right or left, or to the right or left about, etc., forward or backward.

Windage of a *Gun, Mortar,* or *Howitzer.* The difference between the diameter of the bore of the gun, and the diameter of the shot or shell fired from it.

Wings of an Army. The extreme right and left divisions.

Wool Packs. Bags filled with wool, for the purpose of making lodgments in places where there is but little earth to be thrown up to cover the besiegers and working parties from the fire of the garrison.

Words (*Cautionary*). Leading instructions which are given to designate a particular maneuver. The cautionary words precede the words of command, and are issued by the chiefs of corps.

Works. This term implies all the fortifications about a place.

Zigzags. Trenches or paths with several windings, so cut, that the besieged are prevented from enfilading the besieger in his approaches.

Another Romantic Life!